Advances in
Waveform-Agile Sensing
for Tracking

Synthesis Lectures on Algorithms and Software in Engineering

Editor
Andreas Spanias, Arizona State University

Advances in Waveform-Agile Sensing for Tracking
Sandeep Prasad Sira, Antonia Papandreou-Suppappola, and Darryl Morrell
2009

Despeckle Filtering Algorithms and Software for Ultrasound Imaging
Christos P. Loizou and Constantinos S. Pattichis
2008

Advances in Waveform-Agile Sensing for Tracking

Sandeep Prasad Sira, Antonia Papandreou-Suppappola, and Darryl Morrell

ISBN: 978-3-031-00383-7   paperback
ISBN: 978-3-031-01511-3   ebook

DOI 10.1007/978-3-031-02511-3

A Publication in the Springer series
SYNTHESIS LECTURES ON ALGORITHMS AND SOFTWARE IN ENGINEERING

Lecture #2
Series Editors: Andreas Spanias, Arizona State University

Series ISSN
Synthesis Lectures on Algorithms and Software in Engineering
Print 1938-1727   Electronic 1938-1735

Advances in Waveform-Agile Sensing for Tracking

Sandeep Prasad Sira
Zounds Inc., Mesa, AZ

Antonia Papandreou-Suppappola
Arizona State University, Tempe, AZ

Darryl Morrell
Arizona State University at the Polytechnic Campus, Mesa, AZ

SYNTHESIS LECTURES ON ALGORITHMS AND SOFTWARE IN ENGINEERING #2

ABSTRACT

Recent advances in sensor technology and information processing afford a new flexibility in the design of waveforms for agile sensing. Sensors are now developed with the ability to dynamically choose their transmit or receive waveforms in order to optimize an objective cost function. This has exposed a new paradigm of significant performance improvements in active sensing: dynamic waveform adaptation to environment conditions, target structures, or information features.

The manuscript provides a review of recent advances in waveform-agile sensing for target tracking applications. A dynamic waveform selection and configuration scheme is developed for two active sensors that track one or multiple mobile targets. A detailed description of two sequential Monte Carlo algorithms for agile tracking are presented, together with relevant Matlab code and simulation studies, to demonstrate the benefits of dynamic waveform adaptation.

The work will be of interest not only to practitioners of radar and sonar, but also other applications where waveforms can be dynamically designed, such as communications and biosensing.

KEYWORDS

Adaptive waveform selection, waveform-agile sensing, target tracking, particle filtering, sequential Monte Carlo methods, frequency-modulated chirp waveforms.

Contents

CHAPTER 1

Introduction

Waveform diversity is fast becoming one of the most potent methods by which sensing systems can be dynamically adapted to their environment and task to achieve performance gains over nonadaptive systems. Although the development of waveform-agile sensing is relatively new in radar and sonar [1], these features have existed in the bio-sonar of mammals such as bats and dolphins for millions of years. The benefits of such adaptation include improved performance and reduced sensor usage leading to greater system efficiency.

1.1 WAVEFORM-AGILE SENSING

The volume of data gathered by modern sensors often places an overwhelming demand on processing algorithms. To be truly useful, they need to be matched to the sensing objective [2]. For example, increased Doppler resolution may not be much help in identifying an object that is known to be stationary. Therefore, the ability to intelligently direct such sensors to gather the most pertinent data can have significant impact on the performance of a system. As a result, in many applications such as radar, systems are currently being designed with the capability to change the transmit waveform and process it at each time step. In target tracking applications, for example, the advent of waveform-agile sensors, that can shape their transmitted waveforms on-the-fly, makes adaptive waveform configuration schemes possible. These schemes can change the transmitted waveform on a pulse-to-pulse basis to obtain the information that optimally improves the tracker's estimate of the target state. In target detection, on the other hand, especially in challenging scenarios such as in the presence of heavy sea clutter, waveform-adaptation can be exploited to mitigate the effect of the environment and thus improve detection performance.

In general, for active sensing systems, the backscatter from a target depends explicitly on the transmitted waveform. This leads to various modalities where the transmitted waveform can be adapted to differentiate between the target and other reflectors. Thus, polarization-diverse waveforms can be applied to polarimetric radar to yield improved detection and tracking performance while in target recognition applications, the waveforms can be tailored to match the reflective properties of the target.

1.2 WAVEFORM ADAPTATION FOR TRACKING: A REVIEW

Early attempts toward optimization of active target tracking systems treated the sensor and tracking sub-system as completely separate entities [3]. They primarily aimed at improving the matched filter response of the receiver in order to maximize resolution, minimize the effects of mismatching, optimally design the signal for reverberation-limited environments, or for clutter rejection [4]-[6]. Radar signal design was approached via a control theoretic approach in [7, 8] while an information theoretic approach was used in [9]. Waveform design was used to reduce the effect of clutter, ensure track maintenance, and improve detection. For example, in [10], an adaptive pulse-diverse radar/sonar waveform design was proposed to reduce ambiguity due to the presence of strong clutter or jamming signals. In [11] and [12], a wavelet decomposition was used to design waveforms to increase target information extraction in nonstationary environments, while waveform selection for target classification was studied in [13, 14]. The selection of optimal waveform parameters and detection thresholds to minimize track loss in the presence of clutter was investigated in [15, 16].

The primary motivation for the dynamic adaptation of waveforms in tracking applications is that each waveform has different resolution properties, and therefore results in different measurement errors. The choice of waveform can be made such that these errors are small in those dimensions of the target state where the tracker's uncertainty is large while larger errors may be tolerated where the uncertainty is less. Also, these errors are often correlated, and this fact can be exploited to yield variance reduction by conditioning. The first application of this trade-off was presented in [17], where the optimal waveform parameters were derived for tracking one-dimensional target motion using a linear observations model with perfect detection in a clutter-free environment. The tracking was accomplished using a Kalman filter, and waveforms with amplitude-only modulation or linear time-frequency characteristics were used. In this scenario, the problem of selecting the waveform to minimize the tracking error or the validation gate volume can be solved in closed form. In most modern tracking scenarios, however, nonlinear observations models are used, and such closed-form solutions are not possible. The work in [17] was extended to include clutter and imperfect detection, but the linear observations model was still used [18].

Although waveform design optimization was investigated under different performance objectives, the applicability of the results to nonlinear target tracking applications is limited. For example, in [19, 20], the tracking performance of different combinations of waveforms was compared using the expected value of the steady state estimation error. In this work, the authors concluded that the upswept linear frequency-modulated chirp offers very good tracking performance. However, they did not consider dynamic adaptation of waveforms. Recently, the development of waveform libraries for target tracking was studied in [21]. Using an information theoretic criterion and a linear frequency modulated chirp waveform, the authors demonstrated that the maximum expected information about the target state could be obtained by a waveform whose frequency sweep rate was either its minimum or maximum allowable value. This finding, however, does not extend to other performance criteria such as tracking mean square error and observations models, as was demonstrated in [22]. A combined optimization of the detection threshold and the transmitted waveform for target tracking was presented in [23], where a cost function based on the cumulative probability of track loss and the target state covariance was minimized. A strategy to select the optimal sequence of dwells of a phased array radar was developed in [24] by posing the problem as a partially observed Markov decision problem.

Recent work in agile sensing for dynamic waveform selection algorithms for target tracking applications differs from past work [17]-[21, 23, 24], in that it is applicable to tracking scenarios with nonlinear observations models. In addition, it systematically exploits the capabilities of waveforms with varying time-frequency signatures. For example, a configuration algorithm for waveform-agile sensors using generalized frequency-modulated (GFM) chirp signals in nonlinear scenarios was presented in [22] for nonlinear observation scenarios. This work uses a waveform library that is comprised of a number of chirps with different linear or nonlinear instantaneous frequencies. The configuration algorithm simultaneously selects the phase function and configures the duration and frequency modulation (FM) rate of the transmitted waveforms. These waveforms are selected and configured to minimize the predicted mean square error (MSE) in the tracker's estimate of the target state. The MSE is predicted using the Cramér-Rao lower bound (CRLB) on the measurement errors in conjunction with an unscented transform to linearize the observations model.

1.3 ORGANIZATION

This work is organized as follows. In Chapter 2, we provide an overview of target tracking using a particle filter, formulate the waveform-agile tracking problem, and describe the target dynamics and observations models. Chapter 3 develops the waveform selection and configuration algorithm and describes its application to target tracking in narrowband and wideband environments with no clutter and assuming perfect detection. We extend the algorithm to include clutter and imperfect detection in Chapter 4, where applications to the tracking of single and multiple targets are described. We draw conclusions in Chapter 5.

All acronyms used in this work are listed in Table 1.1, while the notation is tabulated in Table 1.2.

Table 1.1: List of acronyms.

Acronym	Description
CRLB	Cramér-Rao lower bound
EFM	Exponential frequency modulated
EKF	Extended Kalman filter
FDSA	Finite difference stochastic approximation
FM	Frequency modulation
GFM	Generalized frequency modulated
HFM	Hyperbolic frequency modulated
LFM	Linear frequency modulated
MC	Monte Carlo
MMSE	Minimum mean square error
MSE	Mean square error
	continued on next page

Table 1.1 – continued from previous page	
Acronym	Description
PFM	Power-law frequency modulated
PRF	Pulse repetition frequency
PRI	Pulse repetition interval
SCR	Signal-to-clutter ratio
SNR	Signal-to-noise ratio
SPSA	Simultaneous perturbation stochastic approximation
UPF	Unscented particle filter
WBAF	Wideband ambiguity function

Table 1.2: Summary of notation.

Notation	Description
$p(x)$	Probability density function of x
$p(x\|y)$	Probability density function of x given y
$E\{x\}$	Expected value of x
$E_{x\|y}\{x\}$	Expected value of x conditioned on y
$\hat{x}$	Estimate of x
x^*	Conjugate of x
A^T	Transpose of A
A^H	Conjugate transpose or Hermitian of A
$AF_{\tilde{s}}(\tau, \nu)$	Narrowband ambiguity function of $\tilde{s}(t)$
E_R	Received signal energy
E_T	Transmitted signal energy
F	State transition matrix
H	Observation matrix
I	Fisher Information Matrix
$J(\cdot)$	Cost function
$N(\theta_k)$	Observation noise covariance matrix as a function of θ_k
N_p	Number of particles
P_d	Probability of detection
P_f	Probability of false alarm
$P_{k\|k}$	Target state covariance matrix at discrete time k given observations up to and including time k
P_{xz}	Cross covariance of the target state and the observation
	continued on next page

Table 1.2 – continued from previous page	
Notation	Description
P_{zz}	Covariance matrix of the observation
Q	Process noise covariance matrix
S	Number of targets in a multiple target scenario
TBW	Time-bandwidth product of the waveform
T_s	Pulse duration
V	Validation gate volume
$\mathbf{V}_k, \mathbf{v}_k$	Observation noise vector at discrete time k
$WAF_{\tilde{s}}(\tau, \sigma)$	Wideband ambiguity function of $\tilde{s}(t)$
$\mathcal{W}$	Weights associated with $\mathcal{X}$ and $\mathcal{Z}$
$\mathbf{W}_k, \mathbf{w}_k$	Process noise vector at discrete time k
$\mathcal{X}$	Sigma points in target space
$\mathbf{X}_k, \mathbf{x}_k$	Target state vector at discrete time k
$\mathcal{Z}$	Sigma points in observation space
$\mathbf{Z}_k, \mathbf{z}_k$	Observation vector at discrete time k
$a(t)$	Amplitude envelope of transmitted signal
b	FM or chirp rate
c	Velocity of propagation
$f(\cdot)$	State transition function
f_s	Sampling frequency
$h(\cdot)$	Observation function
k	Discrete time at which tracker update occurs
$n(t)$	Additive noise component of received signal
q	Process noise intensity
q_2	Information reduction factor
$r(t)$	Received signal
r_k	Target range at discrete time k
$\dot{r}_k$	Target radial velocity or range-rate at discrete time k
$\tilde{s}(t)$	Complex envelope of transmitted signal
$s_R(t)$	Target reflected component of received signal
$s_T(t)$	Transmitted signal
w_k^j	Weight corresponding to the jth particle at discrete time k
x_k, y_k	Target position in Cartesian coordinates at discrete time k
$\dot{x}_k, \dot{y}_k$	Target velocity in Cartesian coordinates at discrete time k
Δ_f	Frequency sweep
ΔT_s	Sampling interval
Λ	Weighting matrix
continued on next page	

Notation	Description
\multicolumn{2}{c}{Table 1.2 – continued from previous page}	
η	Signal to noise ratio
θ_k	Waveform parameter vector at discrete time k
κ	Exponent of phase function in power law chirp
λ	Variance of Gaussian envelope
ν	Doppler shift of received signal
$\xi(t)$	Phase function of the GFM chirp
ρ	Clutter density
σ	Doppler scale
σ_r^2	Variance of r
τ	Delay of received signal
τ_0	Delay corresponding to the target
ψ	Phase function of the PM waveform
ψ_i	Component of ψ corresponding to the ith interval
∇J	Gradient of the cost function

CHAPTER 2

Waveform-Agile Target Tracking Application Formulation

In this chapter, we provide a framework to study the problem of dynamic waveform selection for target tracking. We aim to develop a scheme in which the transmitted waveform is chosen so that it obtains target-related information that minimizes the tracking mean square error (MSE). Since the tracking algorithm controls the choice of transmitted waveform, it forms a critical component of the system. Accordingly, we commence this chapter with a brief overview of target tracking algorithms with special emphasis on the particle filter [25], which has recently gained popularity.

2.1 FILTERING OVERVIEW

Let $\mathbf{X}_k$ represent the state of a system at time k, which we wish to estimate using noisy observations $\mathbf{Z}_k$. We can formulate a general state-space model from the system dynamics and observations models as [26]:

$$\begin{aligned} \mathbf{X}_k &= f_k(\mathbf{X}_{k-1}, \mathbf{W}_{k-1}) \\ \mathbf{Z}_k &= h_k(\mathbf{X}_k, \mathbf{V}_k) \,, \end{aligned} \tag{2.1}$$

where $f_k(\cdot)$ and $h_k(\cdot)$ are possibly nonlinear and time-varying functions of the system state, and $\mathbf{W}_k$ and $\mathbf{V}_k$ are the process and observation noise, respectively. In our chosen scenario, we seek filtered estimates of $\mathbf{X}_k$ based on the set of all available observations $\mathbf{Z}_{1:k} = \{\mathbf{Z}_1, \ldots, \mathbf{Z}_k\}$ and a transmitted waveform parameter sequence $\boldsymbol{\theta}_{1:k} = \{\boldsymbol{\theta}_1, \ldots, \boldsymbol{\theta}_k\}$. From a Bayesian perspective, it is required to construct the probability density function $p(\mathbf{X}_k|\mathbf{Z}_{1:k}, \boldsymbol{\theta}_{1:k})$. Given an initial density $p(\mathbf{X}_0|\mathbf{Z}_0) \equiv p(\mathbf{X}_0)$, we can obtain $p(\mathbf{X}_k|\mathbf{Z}_{1:k}, \boldsymbol{\theta}_{1:k})$ in two stages: prediction and update. The prediction stage uses the system model to first obtain $p(\mathbf{X}_k|\mathbf{Z}_{1:k-1}, \boldsymbol{\theta}_{1:k-1})$ via the Chapman-Kolmogorov equation as [26]

$$p(\mathbf{X}_k|\mathbf{Z}_{1:k-1}, \boldsymbol{\theta}_{1:k-1}) = \int p(\mathbf{X}_k|\mathbf{X}_{k-1}) p(\mathbf{X}_{k-1}|\mathbf{Z}_{1:k-1}, \boldsymbol{\theta}_{1:k-1}) d\mathbf{X}_{k-1} \,. \tag{2.2}$$

At time k, the measurement $\mathbf{Z}_k$ becomes available and it is used to update the predicted estimate via Bayes' rule [26]:

$$p(\mathbf{X}_k|\mathbf{Z}_{1:k}, \boldsymbol{\theta}_{1:k}) = \frac{p(\mathbf{Z}_k|\mathbf{X}_k, \boldsymbol{\theta}_k) p(\mathbf{X}_k|\mathbf{Z}_{1:k-1}, \boldsymbol{\theta}_{1:k-1})}{\int p(\mathbf{Z}_k|\mathbf{X}_k, \boldsymbol{\theta}_k) p(\mathbf{X}_k|\mathbf{Z}_{1:k-1}, \boldsymbol{\theta}_{1:k-1}) d\mathbf{X}_k} \,. \tag{2.3}$$

The recurrence relations in (2.2) and (2.3) define a conceptual solution that, in general, cannot be determined analytically. In the specific case when $f_k(\mathbf{X}_k) = F_k \mathbf{X}_k$ and $h_k(\mathbf{X}_k) = H_k \mathbf{X}_k$, where F_k and H_k are known matrices, the state-space model in (2.1) reduces to a linear model and the Kalman filter [27] is the optimal minimum mean square error (MMSE) estimator. When the state-space model involves nonlinearities, as is often the case, a practical method for implementing the recurrence relations in (2.2) and (2.3) is the particle filter [26].

2.1.1   THE PARTICLE FILTER

In the particle filter, the posterior probability density function $p(\mathbf{X}_k|\mathbf{Z}_{1:k}, \boldsymbol{\theta}_{1:k})$ is represented by a set of N_p random samples and associated weights

$$p(\mathbf{X}_k|\mathbf{Z}_{1:k}, \boldsymbol{\theta}_{1:k}) \approx \sum_{j=1}^{N_p} w_k^j \, \delta(\mathbf{X}_k - \mathbf{X}_k^j) \,, \tag{2.4}$$

where $\mathbf{X}_k^j$ are particles and w_k^j are the corresponding weights. As the number of particles becomes asymptotically large, this representation converges almost surely to the usual functional description of the posterior probability density function.

Since it will not be possible to sample from this density, we introduce an *importance* density $q(\cdot)$ which is easy to sample from. With the samples drawn independently as $\mathbf{X}_k^j \sim q(\mathbf{X}_k|\mathbf{X}_{k-1}^j, \mathbf{Z}_k, \boldsymbol{\theta}_k)$, the weights w_k^j are given by

$$w_k^j \propto w_{k-1}^j \, \frac{p(\mathbf{Z}_k|\mathbf{X}_k^j, \boldsymbol{\theta}_k) p(\mathbf{X}_k^j|\mathbf{X}_{k-1}^j)}{q(\mathbf{X}_k^j|\mathbf{X}_{k-1}^j, \mathbf{Z}_k, \boldsymbol{\theta}_k)} \,. \tag{2.5}$$

A common problem with the particle filter is the degeneracy phenomenon, where, after a few iterations, all but one particle will have negligible weight. This implies that a large computational effort is devoted to updating particles whose contribution to $p(\mathbf{X}_k|\mathbf{Z}_{1:k}, \boldsymbol{\theta}_{1:k})$ is nearly zero. One way to reduce the degeneracy problem is by making a good choice of the importance density. If either the transition density or the likelihood function are very peaked, it will be difficult to obtain samples that lie in regions of high likelihood. This can lead to sample impoverishment and ultimately to the divergence of the filter. The optimal importance density that minimizes the variance of the weights has been shown to be

$$q(\mathbf{X}_k|\mathbf{X}_{k-1}^j, \mathbf{Z}_k, \boldsymbol{\theta}_k)_{opt} = p(\mathbf{X}_k|\mathbf{X}_{k-1}^j, \mathbf{Z}_k, \boldsymbol{\theta}_k) \,.$$

However, it is only possible to compute the optimal importance density for a very limited set of cases [26]. It is often convenient and intuitive to choose the importance density to be the prior

$$q(\mathbf{X}_k^j|\mathbf{X}_{k-1}^j, \mathbf{Z}_k, \boldsymbol{\theta}_k) = p(\mathbf{X}_k^j|\mathbf{X}_{k-1}^j) \,. \tag{2.6}$$

Substitution of (2.6) into (2.5) yields

$$w_k^j \propto w_{k-1}^j \, p(\mathbf{Z}_k | \mathbf{X}_k^j, \boldsymbol{\theta}_k) \, , \tag{2.7}$$

and the estimate $\hat{\mathbf{X}}_k$ is given by

$$\hat{\mathbf{X}}_k = \sum_{j=1}^{N_p} w_k^j \mathbf{X}_k^j \, . \tag{2.8}$$

A second method to avoid degeneracy in the particle filter involves resampling whenever a significant degeneracy is observed. This implies sampling with replacement from the approximate discrete distribution in (2.4) to obtain a new set of N_p particles so that the probability of resampling a particle $\mathbf{X}_k^{j*}$, is w_k^{j*} [26]. This replicates particles with high weights and eliminates those with low weights.

A Matlab code listing for a tutorial example of a particle filter is provided in Appendix C.

2.1.2   THE UNSCENTED PARTICLE FILTER

While the kinematic prior is often chosen as the importance density as in (2.6), its major drawback is that the observation $\mathbf{Z}_k$, which is available when the importance density is sampled, is not used in the sampling process. A sampling scheme that proposes particles in regions of high likelihood can be expected to prevent degeneracy of the filter. One such method is the unscented particle filter (UPF) [28].

The UPF maintains a Gaussian density separately for *each* particle $\mathbf{X}_k^j$ in (2.4). At each sampling instant, each such density is propagated in accordance with the system dynamics and updated with the observation. This process involves the operation of a filter for each density. Since the observations model is nonlinear, the Kalman filter may not be used for this purpose. Accordingly, an unscented Kalman filter, which uses the unscented transform [29] to linearize the observations model, is employed to propagate these densities. Particles are sampled from their corresponding densities and their weights are computed as in (2.5). We will describe the unscented transform in greater detail in Chapter 3.

A Matlab code listing for a tutorial example of an unscented particle filter is provided in Appendix D.

2.2   TRACKING PROBLEM FORMULATION

The tracking algorithms described in Sections 2.1.1 and 2.1.2 are used to track a target that moves in a two-dimensional (2-D) plane (as shown in Figure 2.1). The target is observed using two active, waveform-agile sensors. In this section, we formulate the waveform selection problem in the context of a narrowband, clutter-free environment for a single target. We will extend our formulation to wideband environments and multiple targets in clutter in later chapters.

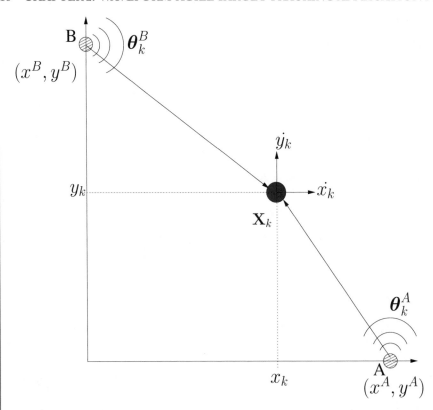

Figure 2.1: Target tracking using two active, waveform-agile sensors. The waveform-agile sensing algorithm selects the next transmitted waveform for each sensor so as to minimize the MSE.

2.2.1   TARGET DYNAMICS

Let $\mathbf{X}_k = [x_k \ y_k \ \dot{x}_k \ \dot{y}_k]^T$ represent the state of a target at time k, where x_k and y_k are the x and y position coordinates, respectively, $\dot{x}_k$ and $\dot{y}_k$ are the respective velocities, and T denotes the vector transpose. The motion of the target is modeled as a Markov process with transition density $p(\mathbf{X}_k|\mathbf{X}_{k-1})$ and initial density $p(\mathbf{X}_0)$ [3]. The target dynamics are modeled by a linear, constant velocity model given by

$$\mathbf{X}_k = F\,\mathbf{X}_{k-1} + \mathbf{W}_k \ . \tag{2.9}$$

The process noise is modeled by the uncorrelated Gaussian sequence $\mathbf{W}_k$. The constant matrix F and the process noise covariance Q are given by

$$F = \begin{bmatrix} 1 & 0 & \Delta T_s & 0 \\ 0 & 1 & 0 & \Delta T_s \\ 0 & 0 & 1 & 0 \\ 0 & 0 & 0 & 1 \end{bmatrix}, \quad Q = q \begin{bmatrix} \frac{\Delta T_s^3}{3} & 0 & \frac{\Delta T_s^2}{2} & 0 \\ 0 & \frac{\Delta T_s^3}{3} & 0 & \frac{\Delta T_s^2}{2} \\ \frac{\Delta T_s^2}{2} & 0 & \Delta T_s & 0 \\ 0 & \frac{\Delta T_s^2}{2} & 0 & \Delta T_s \end{bmatrix}, \quad (2.10)$$

where ΔT_s is the sampling interval and q is a constant.

2.2.2   TRANSMITTED WAVEFORM STRUCTURE

At each sampling epoch k, each sensor transmits a generalized frequency-modulated (GFM) pulse, that is independently chosen from a library of possible waveforms. The transmitted waveform is

$$s_T(t) = \sqrt{2}\text{Re}[\sqrt{E_T}\tilde{s}(t)\exp(j2\pi f_c t)], \quad (2.11)$$

where f_c is the carrier frequency and E_T is the energy of the transmitted pulse. The complex envelope of the transmitted pulse is given by

$$\tilde{s}(t) = a(t)\exp\left(j2\pi b\xi(t/t_r)\right), \quad (2.12)$$

where $a(t)$ is the amplitude envelope, b is a scalar FM rate parameter, $\xi(t/t_r)$ is a real-valued, differentiable phase function, and $t_r > 0$ is a reference time point. We can obtain different FM waveforms with unique time-frequency signatures by varying the phase function $\xi(t/t_r)$, and thus the waveform's instantaneous frequency $\frac{d}{dt}\xi(t/t_r)$. While the linear FM chirp has been popular in radar and sonar [30, 31, 32], nonlinear FM chirps can offer significant advantages [33, 34]. For example, waveforms with hyperbolic instantaneous frequency are Doppler-invariant and are similar to the signals used by bats and dolphins for echolocation [35]. As we will show, one advantage offered by nonlinear GFM chirps is minimal range-Doppler coupling which is an important feature in a tracking system. The waveforms considered in this work are the linear FM (LFM), power-law FM (PFM), hyperbolic FM (HFM) and the exponential FM (EFM) chirps. The amplitude envelope $a(t)$ is not dynamically varied, and is chosen variously to be Gaussian or trapezoidal, as described in the next chapter. In order to ensure a fair comparison among waveforms, we will require that $a(t)$ is chosen so that $\int_{-\infty}^{\infty}|\tilde{s}(t)|^2 dt = 1$.

The waveform transmitted by sensor i, $i = $ A, B, at time k is thus parameterized by the phase function $\xi_k^i(t)$, pulse length λ_k^i, and FM rate b_k^i. We will henceforth assume that $t_r = 1$ in (2.12). Since we seek to dynamically configure the waveforms for both sensors, we define a combined waveform parameter vector for both sensors as $\boldsymbol{\theta}_k = [\boldsymbol{\theta}_k^{A^T} \boldsymbol{\theta}_k^{B^T}]^T$, where $\boldsymbol{\theta}_k^i = [\xi_k^i(t)\ \lambda_k^i\ b_k^i]^T$ is the waveform parameter vector for sensor i.

2.2.3   OBSERVATIONS MODEL

When the transmitted waveform is reflected by the target, its velocity causes a Doppler scaling (compression or dilation) of the time of the complex envelope. The received signal is [36]

$$r(t) = s_R(t) + n(t) ,$$

where

$$s_R(t) = \sqrt{2} \text{Re} \left[\sqrt{E_R} \tilde{s} \left(t - \tau - \frac{2\dot{r}}{c} t \right) \exp \left(j 2\pi f_c (t - \frac{2\dot{r}}{c} t) \right) \right] , \qquad (2.13)$$

and $n(t)$ is additive white Gaussian noise. In (2.13), $\tau = 2r/c$ is the delay of the received signal where r is the range of the target, c is the velocity of propagation, and E_R is the received signal energy. The radial velocity or range-rate of the target with respect to the sensor is $\dot{r}$, and we have used the fact that $\dot{r} \ll c$. From (2.13), we note that the transmitted signal undergoes a Doppler scaling by the factor $1 - (2\dot{r}/c)$ as well as a shift in its carrier frequency. The scaling can be ignored when the time-bandwidth product (TBW) of the waveform satisfies the narrowband condition

$$\text{TBW} \ll \frac{c}{2\dot{r}} , \qquad (2.14)$$

and the Doppler scaling can then be approximated by a simple Doppler shift $v = -2 f_c \dot{r}/c$ [36]. In other words, the approximation says that all frequencies in the signal are shifted equally. An alternative classification of signals as narrowband or wideband is based on the fractional bandwidth, which is the ratio of the bandwidth to the center frequency [37]. When the fractional bandwidth becomes large, the Doppler approximation may not be valid.

For radar, the narrowband condition is easily met as the velocity of propagation is very large ($c \approx 3 \times 10^8$ m/s). When (2.14) holds, the received signal in (2.13) can be approximated by

$$s_R(t) \approx \sqrt{2} \text{Re} \left[\sqrt{E_R} \tilde{s}(t - \tau) \exp \left(j 2\pi (f_c t + vt) \right) \right] . \qquad (2.15)$$

For sonar however, the velocity of propagation is low ($c \approx 1,500$ m/s) and the narrowband approximation in (2.15) may not hold. We will examine this scenario in Chapter 3.

The sensors measure the time delay τ_k^i and Doppler shift v_k^i of the reflected signal. The range of the target is given by $r_k^i = c \tau_k^i / 2$ and its radial velocity is $\dot{r}_k^i = -c v_k^i / 2 f_c$. Let $[r_k^A \ \dot{r}_k^A \ r_k^B \ \dot{r}_k^B]^T$ represent the true range and range-rate with respect to sensors A and B at time k. The errors in the measurement of $[\tau_k^i \ v_k^i]^T$, and eventually in the measurement of $[r_k^i \ \dot{r}_k^i]^T$, depend upon the parameters of the transmitted waveform. These errors are modeled by $\mathbf{V}_k = [\mathbf{v}_k^{A^T} \ \mathbf{v}_k^{B^T}]^T$, a zero-mean, Gaussian noise process with a covariance matrix $N(\boldsymbol{\theta}_k)$. The measurement error covariance matrix changes at each time according to the selected waveform parameters. The nonlinear relation between the target state and its measurement is given by $h^i(\mathbf{X}_k) = [r_k^i \ \dot{r}_k^i]^T$, where

$$\begin{aligned} r_k^i &= \sqrt{(x_k - x^i)^2 + (y_k - y^i)^2} \\ \dot{r}_k^i &= (\dot{x}_k (x_k - x^i) + \dot{y}_k (y_k - y^i))/r_k^i , \end{aligned}$$

and sensor i is located at (x^i, y^i). The measurement originated from the target is given by

$$\mathbf{Z}_k = h(\mathbf{X}_k) + \mathbf{V}_k \, , \tag{2.16}$$

where $h(\mathbf{X}_k) = [h^A(\mathbf{X}_k)^T \, h^B(\mathbf{X}_k)^T]^T$. We assume that the measurement errors $\mathbf{V}_k$ are uncorrelated with the process noise $\mathbf{W}_k$ in (2.9). The state-space model defined by (2.9) and (2.16) is a special case of the general model in (2.1).

2.2.4 MEASUREMENT NOISE COVARIANCE

The impact of a particular waveform on the measurement process appears in the measurement errors in (2.16) which are determined by the resolution properties of the waveform. These properties are reflected in the covariance matrix $N(\boldsymbol{\theta}_k)$ of the process $\mathbf{V}_k$ in (2.16). To derive this covariance, we follow the method developed in [17]. We first assume that the measurement errors at each sensor are independent so that

$$N(\boldsymbol{\theta}_k) = \begin{bmatrix} N(\boldsymbol{\theta}_k^A) & \mathbf{0} \\ \mathbf{0} & N(\boldsymbol{\theta}_k^B) \end{bmatrix} . \tag{2.17}$$

As the form of $N(\boldsymbol{\theta}_k^A)$ and $N(\boldsymbol{\theta}_k^B)$ is similar, in the characterization that follows, we consider the noise covariance $N(\boldsymbol{\theta}_k^i)$ for one sensor only.

The estimation of delay and Doppler at each sensor is performed by matched filters that correlate the received waveform with delayed and Doppler-shifted versions of the transmitted waveform. The peak of the correlation function is a maximum likelihood estimator of the delay and Doppler of the received waveform. The magnitude of the correlation between a waveform and time-frequency shifted replicas of itself is given by the ambiguity function, which provides a well-established starting point for the evaluation of the effects of a waveform on the measurement error. While considering a narrowband environment, we use the narrowband ambiguity function that is defined as [36]

$$AF_{\tilde{s}}(\tau, \nu) = \int_{-\infty}^{\infty} \tilde{s}\left(t + \frac{\tau}{2}\right) \tilde{s}^*\left(t - \frac{\tau}{2}\right) \exp(-j2\pi\nu\,(t)dt \, . \tag{2.18}$$

When the narrowband approximation in (2.14) does not hold, the narrowband ambiguity function has to be replaced with its wideband version (see Chapter 3). The Cramér-Rao lower bound (CRLB) of the matched filter estimator can be obtained by inverting the Fisher information matrix which is the Hessian of the ambiguity function in (2.18), evaluated at the true target delay and Doppler [36]. Equivalently, the elements of the Fisher information matrix can also be determined directly from the complex envelope of the waveform [38, 39]. For the time-varying GFM chirps, we provide this computation and the resulting $N(\boldsymbol{\theta}_k)$ in Appendices A and B.

It is important to note that the CRLB depends only on the properties of the ambiguity function at the origin. The location of the peak of the ambiguity function in the delay-Doppler plane is affected by its sidelobes which are not considered in the evaluation of the CRLB, thus limiting its effectiveness as a measure of the waveform's estimation performance. Another method of computing the measurement error covariance with explicit dependence on the sidelobes was proposed in [19]. This method is based on the notion of a resolution cell that encloses the ambiguity function contour at a given probability of detection with the true target location assumed to be distributed uniformly within the cell. However, the size of the resolution cell and the associated measurement error covariance increases with the probability of detection and thus the signal-to-noise ratio (SNR). It is therefore not feasible for use in an adaptive scheme. If high SNR is assumed, the sidelobes of the ambiguity function may be neglected and the estimator can be assumed to achieve the CRLB. We make this assumption and set $N(\boldsymbol{\theta}_k^i)$ equal to the CRLB, which is computed using the waveform in (2.12) with parameter $\boldsymbol{\theta}_k^i$.

2.2.5   WAVEFORM SELECTION PROBLEM STATEMENT

The criterion we use for the dynamic waveform selection is the minimization of the tracking MSE. When this selection is carried out, we do not have access to either the observation or the target state at the next sampling instant. The predicted squared error is thus a random variable, and we seek to minimize its expected value. We therefore attempt to minimize the cost function

$$J(\boldsymbol{\theta}_k) = E_{\mathbf{X}_k, \mathbf{Z}_k | \mathbf{Z}_{1:k-1}} \left\{ (\mathbf{X}_k - \hat{\mathbf{X}}_k)^T \boldsymbol{\Lambda} (\mathbf{X}_k - \hat{\mathbf{X}}_k) \right\} \qquad (2.19)$$

over the space of allowable waveforms with parameter vectors $\boldsymbol{\theta}_k$. Here, $E\{\cdot\}$ is an expectation over predicted states and observations, $\boldsymbol{\Lambda}$ is a weighting matrix that ensures that the units of the cost function are consistent, and $\hat{\mathbf{X}}_k$ is the estimate of $\mathbf{X}_k$ given the sequence of observations $\mathbf{Z}_{1:k}$. Note that the cost in (2.19) results in a one-step ahead or myopic optimization. Although it is possible to formulate a nonmyopic (multi-steps ahead) cost function, the computational complexity associated with its minimization grows with the horizon of interest. The cost in (2.19) cannot be computed in closed form due to the nonlinear relationship between the target state and the measurement. In Chapter 3, we will present two methods of approximating it and obtaining the waveform that yields the lowest approximate cost.

CHAPTER 3

Dynamic Waveform Selection with Application to Narrowband and Wideband Environments

The waveform selection algorithm seeks to choose a waveform that minimizes the mean square error (MSE) or the expected cost in (2.19). A key element of the algorithm, therefore, is the computation of the expected cost corresponding to each potential waveform. In this chapter, we explore the difficulty of computing the predicted MSE in a nonlinear observation environment and present two methods to approximate it. A waveform selection procedure that utilizes these methods is developed and applied to the tracking of a single target in narrowband and wideband environments.

3.1 PREDICTION OF THE MSE

The waveform selection is guided by the objective of minimizing the tracking MSE. Due to the nonlinear observations model, however, the cost function $J(\boldsymbol{\theta}_k)$ in (2.19) cannot be evaluated in closed form. This is in significant contrast with previous work on waveform optimization problems such as [17]-[21]. When linear observations and target dynamics models are used, the Kalman filter [27] is the minimum mean square error (MMSE) estimator of the target state given the sequence of observations. In such situations the covariance update of the Kalman filter provides a means of obtaining the cost of using a particular waveform in closed form. This fact was exploited in [17] to determine the optimal waveform selection for the minimization of two cost functions: tracking MSE and validation gate volume. In the measurement scenario described in Section 2.2.3, the optimal estimator of the target state cannot be determined in closed form. This precludes the determination of the cost function in closed form. We must therefore attempt to approximate it. Next, we present two approaches to this problem, the first of which is based on Monte Carlo (MC) methods, while the second employs the unscented transform.

3.2 STOCHASTIC APPROXIMATION

Recognizing that the expected cost in (2.19) is an integral over the joint distribution of the target state $\mathbf{X}_k$ and the observation $\mathbf{Z}_k$, we may approximate it by evaluating the integral using MC methods. The expectation in (2.19) can be expanded as

$$J_k(\boldsymbol{\theta}_k) = \int \int (\mathbf{X}_k - \hat{\mathbf{X}}_k)^T \boldsymbol{\Lambda} (\mathbf{X}_k - \hat{\mathbf{X}}_k) \, p(\mathbf{Z}_k|\mathbf{X}_k, \boldsymbol{\theta}_k) \, p(\mathbf{X}_k|\mathbf{Z}_{1:k-1}, \boldsymbol{\theta}_{1:k-1}) \, d\mathbf{X}_k \, d\mathbf{Z}_k . \quad (3.1)$$

It can be approximated for large N_x and N_z as

$$J_k(\boldsymbol{\theta}_k) \approx \frac{1}{N_x} \sum_{n=1}^{N_x} \frac{1}{N_z} \sum_{p=1}^{N_z} \left(\tilde{\mathbf{X}}_n - \hat{\mathbf{X}}_{n|\tilde{\mathbf{Z}}_p, \mathbf{Z}_{1:k-1}} \right)^T \boldsymbol{\Lambda} \left(\tilde{\mathbf{X}}_n - \hat{\mathbf{X}}_{n|\tilde{\mathbf{Z}}_p, \mathbf{Z}_{1:k-1}} \right) , \qquad (3.2)$$

where $\tilde{\mathbf{X}}_n$, $n = 1, \ldots, N_x$, are independent samples of predicted states drawn from the estimate of the density $p(\mathbf{X}_k | \mathbf{Z}_{1:k-1}, \boldsymbol{\theta}_{1:k-1})$, which is obtained from the tracker, and $\tilde{\mathbf{Z}}_p$, $p = 1, \ldots, N_z$, are predicted observations drawn independently from the likelihood $p(\mathbf{Z}_k | \tilde{\mathbf{X}}_n, \boldsymbol{\theta}_k)$. The estimate of $\tilde{\mathbf{X}}_n$, given the observations $\mathbf{Z}_{1:k-1}$ and $\tilde{\mathbf{Z}}_p$ is denoted as $\hat{\mathbf{X}}_{n|\tilde{\mathbf{Z}}_p, \mathbf{Z}_{1:k-1}}$ and may be computed by a secondary particle filter. The search for the waveform parameter that minimizes (3.2) can be accomplished by an iterative stochastic steepest-descent method. To implement this method, we must first obtain the gradient of the expected cost.

3.2.1   CALCULATION OF THE GRADIENT
The gradient of the cost function in (2.19) is

$$\nabla J_k(\boldsymbol{\theta}_k) = \begin{bmatrix} \dfrac{\partial J_k(\boldsymbol{\theta}_k)}{\partial \boldsymbol{\theta}_k^A} \\ \dfrac{\partial J_k(\boldsymbol{\theta}_k)}{\partial \boldsymbol{\theta}_k^B} \end{bmatrix} .$$

If the conditions for the interchange of the derivative and expectation [40] in (3.1) are satisfied, it is possible to directly estimate the gradient. The Likelihood Ratio or the Score Function Method [41] provide a convenient means for such an estimation. However, the estimate $\hat{\mathbf{X}}_k$ is a function of the observations $\mathbf{Z}_{1:k}$, which depend on $\boldsymbol{\theta}_k$. Thus, it is not possible to evaluate in closed form the partial derivatives

$$\frac{\partial \left((\mathbf{X}_k - \hat{\mathbf{X}}_k)^T \boldsymbol{\Lambda} (\mathbf{X}_k - \hat{\mathbf{X}}_k) \right)}{\partial \boldsymbol{\theta}_k^A} \quad \text{and} \quad \frac{\partial \left((\mathbf{X}_k - \hat{\mathbf{X}}_k)^T \boldsymbol{\Lambda} (\mathbf{X}_k - \hat{\mathbf{X}}_k) \right)}{\partial \boldsymbol{\theta}_k^B} ,$$

and, as a result, direct gradient approximation methods cannot be used. We therefore use the Simultaneous Perturbation Stochastic Approximation (SPSA) method [42] to approximate the gradient.

3.2.2   SIMULTANEOUS PERTURBATION STOCHASTIC APPROXIMATION
The Kiefer-Wolfowitz finite difference stochastic approximation (FDSA) algorithm [43] provides a solution to the calculation of the gradient of cost functions where the exact relationship between the parameters being optimized and the cost function cannot be determined. In this algorithm, each element of the multivariate parameter is varied in turn by a small amount, and the cost function is evaluated at each new value of the parameter. The gradient is then approximated by determining the rate of change of the cost function over the perturbation in the parameters. As the algorithm iterates, the perturbations asymptotically tend to zero and the estimate of the gradient converges to the true

value. This method suffers from a high computational load in that the cost function must be evaluated $d + 1$ times for a d-dimensional parameter. The SPSA method, on the other hand, perturbs all the parameters simultaneously and requires only two evaluations of the cost function at each iteration. It was proved in [42] that under reasonably general conditions, SPSA achieves the same level of statistical accuracy for a given number of iterations and uses d times fewer function evaluations. We use the gradient approximation in a steepest descent algorithm to iteratively determine the parameters of the waveform $\boldsymbol{\theta}_k \in \Re^d$, that minimize the cost function. To describe the algorithm, let $\boldsymbol{\theta}^l$ denote the waveform parameter at the lth iteration in a stochastic gradient descent algorithm,

$$\boldsymbol{\theta}^{l+1} = \pi_\Theta \left(\boldsymbol{\theta}^l - a_l \widehat{\nabla J}_l(\boldsymbol{\theta}^l) \right) , \tag{3.3}$$

where the scalar sequence a_l satisfies the properties

$$\sum_{l=1}^{\infty} a_l = \infty \quad \text{and} \quad \sum_{l=1}^{\infty} a_l^2 < \infty . \tag{3.4}$$

In (3.3), $\pi_\Theta(\cdot)$ is a projection operator that constrains $\boldsymbol{\theta}^{l+1}$ to lie within a given set of values which represent practical limitations on the waveform parameters [44], and $\widehat{\nabla J}_l(\boldsymbol{\theta}^l)$ is the estimate of the gradient of the cost function at parameter $\boldsymbol{\theta}^l$.

Let $\boldsymbol{\Delta}_l \in \Re^d$ be a vector of d independent zero-mean random variables $\{\Delta_{l1}, \Delta_{l2}, \ldots, \Delta_{ld}\}$ such that $E\{\Delta_{li}^{-1}\}$ is bounded. This precludes Δ_{li} from being uniformly or normally distributed. The authors in [42] suggest that $\boldsymbol{\Delta}_l$ can be symmetrically Bernoulli distributed. Let c_l be a scalar sequence such that

$$c_l > 0 \quad \text{and} \quad \sum_{l=1}^{\infty} \left(\frac{a_l}{c_l} \right)^2 < \infty . \tag{3.5}$$

The parameter $\boldsymbol{\theta}^l$ is simultaneously perturbed to obtain $\boldsymbol{\theta}^{l(\pm)} = \boldsymbol{\theta}^l \pm c_l \boldsymbol{\Delta}_l$. Let

$$\hat{J}_l^{(+)} = \hat{J}_l(\boldsymbol{\theta}^l + c_l \boldsymbol{\Delta}_l) \quad \text{and} \quad \hat{J}_l^{(-)} = \hat{J}_l(\boldsymbol{\theta}^l - c_l \boldsymbol{\Delta}_l) ,$$

represent noisy measurements of the cost function evaluated using the perturbed parameters $\boldsymbol{\theta}^{l(+)}$ and $\boldsymbol{\theta}^{l(-)}$. The estimate of the gradient at the lth iteration is then given by

$$\widehat{\nabla J}_l(\boldsymbol{\theta}^l) = \begin{bmatrix} \dfrac{\hat{J}_l^{(+)} - \hat{J}_l^{(-)}}{2 c_l \Delta_{l1}} \\ \vdots \\ \dfrac{\hat{J}_l^{(+)} - \hat{J}_l^{(-)}}{2 c_l \Delta_{ld}} \end{bmatrix} . \tag{3.6}$$

Even when d is large, the simultaneous perturbation ensures that no more than two evaluations of the cost function are required at each iteration, thus reducing the computational cost.

In many problems of interest, the components of the gradient have significantly different magnitudes. In these problems, better convergence of the gradient descent algorithm is obtained by using different values of a_l and c_l for each component of the gradient.

3.2.3   STOCHASTIC GRADIENT DESCENT ALGORITHM

At time $k - 1$, the tracking particle filter provides an estimate of the posterior probability density function $p(\mathbf{X}_{k-1}|\mathbf{Z}_{1:k-1}, \boldsymbol{\theta}_{1:k-1})$ of the target state given the observations up to time $k - 1$. The stochastic gradient descent algorithm returns a value of $\boldsymbol{\theta}_k$ which minimizes the expected squared tracking error at time k. The operation of the steepest descent algorithm with the SPSA-based gradient approximation is described in the following algorithm [45].

Algorithm 1: Stochastic Optimization

Initialization

- We choose $\boldsymbol{\theta}^{(0)}$ as the initial value of an iterative search for the parameter $\boldsymbol{\theta}_k$. $\boldsymbol{\theta}^{(0)}$ is chosen so that the pulse length is set to the middle of the allowed range and the FM rate is set to 0, since it can take positive or negative values.

- The sequences a_l and c_l in (3.3) and (3.5) are selected.

- We set the number of iterations to L. In our simulations, we have found that $L = 500$ provides satisfactory results. However, the choice of a stopping time for a stochastic gradient descent algorithm is not trivial and needs to be further investigated.

- The collection of particles, $\mathbf{X}_{k-1}^j$ and weights w_{k-1}^j that represent the estimated density $p(\mathbf{X}_{k-1}|\mathbf{Z}_{1:k-1}, \boldsymbol{\theta}_{1:k-1})$, are projected forward using the state dynamics in (2.9) to obtain $p(\mathbf{X}_k|\mathbf{Z}_{1:k-1}, \boldsymbol{\theta}_{1:k-1})$, which is an estimate of the probability density function of the state at time k, given the observations up to time $k - 1$.

Iteration $l = 1 : L$

- Obtain an estimate of the gradient $\widehat{\nabla J}_{(l-1)}(\boldsymbol{\theta}^{(l-1)})$ according to Algorithm 2, described below.

- Calculate the value of the waveform parameter $\boldsymbol{\theta}^l$ to be used in the next iteration according to (3.3).

After the gradient descent algorithm completes L iterations, $\boldsymbol{\theta}^L$ is the selected waveform parameter and $\boldsymbol{\theta}_k = \boldsymbol{\theta}^L$.

Algorithm 2: Gradient Estimation

Initialize

- We select the values of N_x and N_z in (3.2). In our simulations $N_x = N_z = 10$.

- $\boldsymbol{\Delta}_l$ is chosen such that each element is a ± 1 Bernoulli distributed random variable.

- We obtain the perturbed parameters $\boldsymbol{\theta}^{l(\pm)} = \boldsymbol{\theta}^l \pm c_l \boldsymbol{\Delta}_l$.

Iterate for $\boldsymbol{\theta}^{l*} = \boldsymbol{\theta}^{l(+)}, \boldsymbol{\theta}^{l(-)}$

- To obtain the expectation of the predicted mean square tracking error over future states and observations, we first propose N_x states, $\tilde{\mathbf{X}}_k^n, n = 1, \ldots, N_x$, by sampling the density $p(\mathbf{X}_k | \mathbf{Z}_{1:k-1}, \boldsymbol{\theta}_{1:k-1})$. In practice, this density is simply a set of particles and the sampling is achieved by selecting N_x particles.

- For each of the N_x proposed states, we choose N_z possible observations, $\tilde{\mathbf{Z}}_k^p, p = 1, \ldots, N_z$, by sampling the density $p(\mathbf{Z}_k | \tilde{\mathbf{X}}_k^n, \boldsymbol{\theta}^{l*})$. The waveform parameter, which is the perturbed parameter from the current iteration of Algorithm 1, parameterizes this density and determines the measurement errors. This accordingly changes the measurement noise covariance matrix at each iteration.

- For each of the N_x predicted states, N_z estimates of the state are computed using a secondary particle filter. The particles $\mathbf{X}_k^j, j = 1, \ldots, N_p$ are already available as samples from $p(\mathbf{X}_k | \mathbf{Z}_{1:k-1}, \boldsymbol{\theta}_{1:k-1})$. The weights w_k^j for the pth estimate are calculated according to the likelihood $p(\tilde{\mathbf{Z}}_k^p | \mathbf{X}_k^j, \boldsymbol{\theta}^{l*})$ and the estimate is given by $\sum_{j=1}^{N_p} w_k^j \mathbf{X}_k^j$. For each estimate the total squared tracking error is calculated as $(\tilde{\mathbf{X}}_k^n - \hat{\mathbf{X}}_{n|\tilde{\mathbf{Z}}_p, \mathbf{Z}_{1:k-1}})^T \Lambda (\tilde{\mathbf{X}}_k^n - \hat{\mathbf{X}}_{n|\tilde{\mathbf{Z}}_p, \mathbf{Z}_{1:k-1}})$.

- After $N_x N_z$ such iterations we calculate the average of the squared tracking error to obtain $\hat{J}_l^*$.

We obtain the gradient estimate $\widehat{\nabla J_l}(\boldsymbol{\theta}^l)$ from $\hat{J}_l^{(+)}$ and $\hat{J}_l^{(-)}$ according to (3.6).

3.2.4   DRAWBACKS

This method was applied to the problem of configuring the sensors when the waveform library consisted of the linear frequency modulated (LFM) chirp alone [45]. It does not appear feasible to extend the method to cases where the library contains a large number of candidate waveforms because it suffers from two drawbacks. Firstly, the MC approximation is computationally intensive. This is due to the fact that each sample in the average in (3.2) requires an estimate of the predicted state $\hat{\mathbf{X}}_{n|\tilde{\mathbf{Z}}_p, \mathbf{Z}_{1:k-1}}$ in (3.2), which is made by a particle filter. In addition, each iteration of the stochastic steepest-descent algorithm requires at least two evaluations of the cost function to calculate the gradient. Typically, however, averaging is required to reduce the noise in the gradient estimate, and this further increases the computational burden. Secondly, a large number of iterations is required to achieve convergence due to the noise that accompanies each gradient estimate. The available range of the duration parameter is small, and this poses further challenges in achieving an accurate solution. In order to overcome these difficulties, we propose a more computationally efficient method based on the unscented transform.

3.3   UNSCENTED TRANSFORM BASED APPROXIMATION

The Kalman filter covariance update equation provides a mechanism to recursively compute the covariance of the state estimate provided the observations and dynamics models are linear [46].

The observations model in (2.16) does not satisfy this requirement but it can be linearized and this approach can still be applied. The standard approach to this problem is the extended Kalman filter (EKF) which approximates the nonlinearity in (2.16) by a Taylor series expansion about a nominal target state and discards the higher order terms. An improvement on this method is the unscented transform [29]. The resulting unscented Kalman filter assumes that the density of the state given the observations is Gaussian and employs the unscented transform to compute its statistics under a nonlinear transformation. This approach has been shown to outperform the EKF [29]. Thus, we use the covariance update of the unscented Kalman filter to approximate the cost function as follows.

Let $P_{k-1|k-1}$ represent the covariance of the state estimate at time $k - 1$. We wish to approximate the covariance $P_{k|k}(\boldsymbol{\theta}_k)$ that would be obtained if a waveform characterized by its parameter vector $\boldsymbol{\theta}_k$ was used to obtain a measurement at time k. First, the dynamics model in (2.9) is used to obtain the predicted mean and covariance as

$$\hat{\mathbf{X}}_{k|k-1} = F\hat{\mathbf{X}}_{k-1|k-1} \quad \text{and} \quad P_{k|k-1} = F P_{k-1|k-1} F^{\mathrm{T}} + Q ,$$

respectively. Next, we select $2N + 1$ sigma points $\boldsymbol{\mathcal{X}}_n$, and corresponding weights $\mathcal{W}_n$ as [29]

$$
\begin{aligned}
\boldsymbol{\mathcal{X}}_0 &= \hat{\mathbf{X}}_{k|k-1}, & \mathcal{W}_0 &= \beta/(n+\beta) , \\
\boldsymbol{\mathcal{X}}_n &= \hat{\mathbf{X}}_{k|k-1} + \left(\sqrt{(n+\beta)P_{k|k-1}}\right)_n , & \mathcal{W}_n &= 1/(2(n+\beta)) , \\
\boldsymbol{\mathcal{X}}_{N+n} &= \hat{\mathbf{X}}_{k|k-1} - \left(\sqrt{(n+\beta)P_{k|k-1}}\right)_n , & \mathcal{W}_{N+n} &= 1/(2(n+\beta)) ,
\end{aligned}
\tag{3.7}
$$

where $\beta \in \Re$ is an appropriately chosen scalar [29], and $\left(\sqrt{(n+\beta)P_{k|k-1}}\right)_n$ is the nth row or column of the matrix square root of $(n+\beta)P_{k|k-1}$.

A transformed set of sigma points $\boldsymbol{\mathcal{Z}}_n = h(\boldsymbol{\mathcal{X}}_n)$ is computed. Then, we calculate the covariances

$$P_{zz} = \sum_{n=0}^{2N+1} \mathcal{W}_n \left(\boldsymbol{\mathcal{Z}}_n - \bar{\boldsymbol{\mathcal{Z}}}\right)\left(\boldsymbol{\mathcal{Z}}_n - \bar{\boldsymbol{\mathcal{Z}}}\right)^T , \tag{3.8}$$

$$P_{xz} = \sum_{n=0}^{2N+1} \mathcal{W}_n \left(\boldsymbol{\mathcal{X}}_n - \bar{\boldsymbol{\mathcal{X}}}\right)\left(\boldsymbol{\mathcal{Z}}_n - \bar{\boldsymbol{\mathcal{Z}}}\right)^T , \tag{3.9}$$

where

$$\bar{\boldsymbol{\mathcal{Z}}} = \sum_{n=0}^{2N+1} \mathcal{W}_n \boldsymbol{\mathcal{Z}}_n \quad \text{and} \quad \bar{\boldsymbol{\mathcal{X}}} = \sum_{n=0}^{2N+1} \mathcal{W}_n \boldsymbol{\mathcal{X}}_n .$$

The estimate of the updated covariance that would result if a waveform with parameter $\boldsymbol{\theta}_k$, and hence a measurement error covariance $N(\boldsymbol{\theta}_k)$, were used is then obtained as

$$P_{k|k}(\boldsymbol{\theta}_k) \approx P_{k|k-1} - P_{xz}(P_{zz} + (N(\boldsymbol{\theta}_k))^{-1} P_{xz}{}^T . \tag{3.10}$$

The approximate cost $J(\boldsymbol{\theta}_k)$ in (2.19) is then computed as the trace of $\boldsymbol{\Lambda} P_{k|k}(\boldsymbol{\theta}_k)$. Note that the matrices $P_{k|k-1}$, P_{xz} and P_{zz} in (3.10) have to be calculated only once for each sampling interval. The measurement noise covariance associated with each candidate waveform can be computed and (3.10) can be repeatedly used to obtain the predicted cost. Since the matrices $N(\boldsymbol{\theta}_k)$ can be computed offline, this method is suitable for online implementation. The application of this approximation to waveform selection is described next.

3.4   ALGORITHM FOR WAVEFORM SELECTION

The overall configuration and tracking algorithm is shown in Figure 3.1. The dynamic waveform selection is performed by a search over the space of allowable waveforms for the candidate that results in the lowest cost, which is then chosen as the sensor configuration for the next sampling instant. While gradient-based methods could be used to optimize the duration and FM rate, which are continuous parameters, we chose to use a rectangular grid search approach over a finite set of values since it is computationally cheaper and works well in practice. For each sensor, R grid points for λ and L grid points for the frequency sweep Δ_f are evenly spaced over the intervals $[\lambda_{min}, \lambda_{max}]$ and $[0, \Delta_f^{max}]$, respectively. Here, λ_{min} and λ_{max} are bounds that are determined by the constraints on the pulse duration, and Δ_f^{max} is the maximum allowed frequency sweep. The values of the FM rates b, for each λ, corresponding to the phase function $\xi(t)$, are calculated for each frequency sweep Δ_f. We also consider the upsweep and downsweep of the frequency for each configuration which leads to $2RL$ possible configurations per sensor for each phase function $\xi(t)$. The grid for a single phase function thus contains $(2RL)^2$ points.

The expected cost is computed at each grid point using the procedure described in Section 3.3. The values of duration and frequency sweep that minimize the expected cost form the center of a new grid whose boundaries are taken as the immediate surrounding grid points. The expected cost is computed at each point on the new grid, and this procedure is repeated several times to find the configuration for each phase function that minimizes the expected cost. All combinations of these phase-specific configurations are now tested to determine the one that yields the lowest cost across all phase functions.

3.5   NARROWBAND ENVIRONMENT

In this section, we apply the waveform selection and configuration algorithm to target tracking in narrowband environments, where perfect detection and an absence of clutter are assumed. This implies that the narrowband condition in (2.14) is assumed to be satisfied. We begin by describing the evaluation of the CRLB for generalized frequency modulated (GFM) pulses, and then apply it to a simulation study [47].

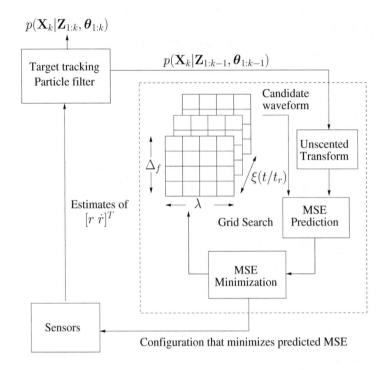

Figure 3.1: Block diagram of the waveform selection and configuration algorithm.

3.5.1   WAVEFORM STRUCTURE

In this study, we use the waveform defined in (2.12) with a Gaussian envelope defined as

$$a(t) = \left(\frac{1}{\pi\lambda^2}\right)^{\frac{1}{4}} \exp\left(-\frac{(t/t_r)^2}{2\lambda^2}\right),\tag{3.11}$$

where λ is treated as a duration parameter. Although we consider the pulse to be of infinite duration for the calculation of the CRLB that follows, the effective pulse length, T_s, is chosen to be the time interval over which the signal amplitude is greater than 0.1% of its maximum value. This further determines the value of $\lambda = T_s/\alpha$, where $\alpha = 7.4338$ [17]. It can be shown that the resulting difference in the CRLB computation is small.

The phase functions used in this simulation example result in the linear FM (LFM), hyperbolic FM (HFM,) and the exponential FM (EFM) chirps. The phase function definitions and the resulting frequency sweep are shown in Table 3.1.

Table 3.1: Phase function and bandwidth of GFM waveforms with Gaussian envelopes.

Waveform	Phase Function, $\xi(t)$	Frequency sweep, Δ_f
LFM	t^2	bT_s
HFM	$\ln(T + |t|), T > 0$	b/T
EFM	$\exp(|t|)$	$b\exp(T_s/2)$

3.5.2 CRLB FOR GFM PULSES

As described in Section 2.2.4, the negative of the second derivatives of the ambiguity function, evaluated at $\tau = 0$, $\nu = 0$, yield the elements of the Fisher information matrix [36]. Denoting the signal-to-noise ratio (SNR) by η, the Fisher information matrix for the GFM waveforms defined by (2.12) and (3.11) is

$$I = \eta \begin{bmatrix} \frac{1}{2\lambda^2} + g(\xi) & 2\pi f(\xi) \\ 2\pi f(\xi) & (2\pi)^2 \frac{\lambda^2}{2} \end{bmatrix} .$$

We computed its elements as

$$-\frac{\partial^2 A F_{\tilde{s}}(\tau, \nu)}{\partial \tau^2}\bigg|_{\substack{\tau=0 \\ \nu=0}} = \frac{1}{2\lambda^2} + g(\xi) ,$$

$$-\frac{\partial^2 A F_{\tilde{s}}(\tau, \nu)}{\partial \tau \partial \nu}\bigg|_{\substack{\tau=0 \\ \nu=0}} = 2\pi f(\xi) ,$$

$$-\frac{\partial^2 A F_{\tilde{s}}(\tau, \nu)}{\partial \nu^2}\bigg|_{\substack{\tau=0 \\ \nu=0}} = (2\pi)^2 \frac{\lambda^2}{2} ,$$

where (see Appendix A)

$$g(\xi) = (2\pi b)^2 \int_{-\infty}^{\infty} \frac{1}{\lambda\sqrt{\pi}} \exp\left(-\frac{t^2}{\lambda^2}\right) [\xi'(t)]^2 dt , \qquad (3.12)$$

$$f(\xi) = 2\pi b \int_{-\infty}^{\infty} \frac{t}{\lambda\sqrt{\pi}} \exp\left(-\frac{t^2}{\lambda^2}\right) \xi'(t) dt , \qquad (3.13)$$

and $\xi'(t) = d\xi(t)/dt$. The CRLB on the variance of the error in the estimate of $[\tau, \nu]^T$ is given by I^{-1}.

Since $r = c\tau/2$ and $\dot{r} = -c\nu/(2f_c)$, the CRLB on the error variance of the estimate of $[r, \dot{r}]^T$ is given by $\Gamma I^{-1} \Gamma^T$ where $\Gamma = \text{diag}(c/2, c/(2f_c))$. Note that I^{-1} depends explicitly on the waveform parameters due to (3.12) and (3.13). The measurement error covariance at the ith sensor is $N(\boldsymbol{\theta}_k^i) = \frac{1}{\eta_k^i}\Gamma(I_k^i)^{-1}\Gamma^T$, where η_k^i is the SNR at sensor i. Since we assume that the noise at each sensor is independent, $N(\boldsymbol{\theta}_k)$ is given by (2.17).

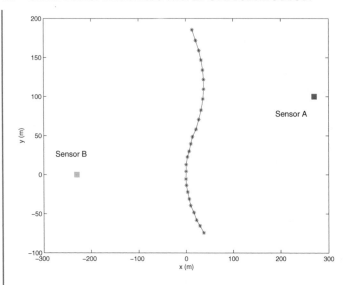

Figure 3.2: Trajectory of the target motion.

3.5.3   SIMULATION

In this simulation example, we consider two fixed, waveform-agile sensors tracking a single under-water target as it moves in two dimensions. In the first example, we consider the dynamic selection of the duration and FM rate of the LFM chirp alone, while in the second example, we adapt the phase function as well.

The carrier frequency was $f_c = 25$ kHz and the velocity of sound in water was assumed to be a constant 1,500 m/s. The trajectory of the target is shown in Figure 3.2 and the sensors A and B were located at (275, 100) m and (-225,0) m, respectively. The SNR at the ith sensor was modeled according to

$$\eta_k^i = \left(\frac{r_0}{r_k^i}\right)^4, \tag{3.14}$$

where r_0 is the range at which an SNR of 0 dB was obtained. In this example, $r_0 = 5,000$ m. The pulse duration was constrained to be $T_s \in [0.01, 0.3]$ s so that $\lambda \in [1.3, 40.4]$ ms, while the FM rate was $b \in [0, 10]$ kHz/s. In order to ensure that the narrowband condition (2.14) was satisfied, we chose the maximum frequency sweep as $\Delta_f^{max} = 100$ Hz. The sampling interval was $\Delta T_s = 2$ s while the process noise intensity in (2.10) was $q = 1$. We used the UPF as the target tracker and the weighting matrix in (2.19) was set to $\Lambda = \text{diag}[1, 1, 4\,\text{s}^2, 4\,\text{s}^2]$ so that the cost was in units of m^2. All results were averaged over 500 simulation runs.

Example 1: LFM Only

We first test the waveform parameter selection algorithm by choosing λ and b for each sensor when the library of waveforms consists of the LFM chirp only. For comparison, we also investigate the performance of the tracking algorithm when the sensors do not dynamically adapt their transmitted waveforms. The fixed waveforms in this case correspond to an LFM chirp with the minimum possible duration $\lambda = 1.3$ ms and maximum allowed frequency sweep, and the LFM chirp with the maximum allowed duration $\lambda = 40.4$ ms and maximum allowed frequency sweep. Note that the latter configuration corresponds to the case of the maximum time-bandwidth product, which in conventional radar and sonar literature, is often considered to be the best choice for tracking applications.

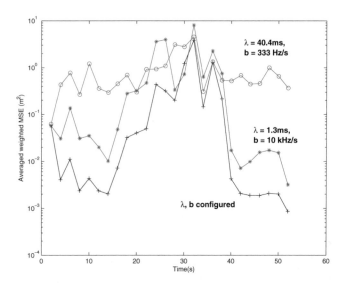

Figure 3.3: Averaged MSE when the sensors dynamically configure LFM chirp waveforms only.

Figure 3.3 shows the averaged MSE obtained when fixed as well as dynamically configured LFM waveforms are used by the sensors. We observe that the configured waveform offers significantly better performance over the fixed waveform with the maximum time-bandwidth product. The increase in the MSE at $k = 15$ (30 (s)) is due to the fact that, at this time, the target crosses the line joining the sensors. This results in large tracking errors leading to large MSE. The response of the algorithm to this situation is seen in Figure 3.4, where the selected pulse length and FM rate are shown for each sensor. We note that the sensors dynamically change the waveform parameters as the tracking progresses. Specifically, at $k = 15$ (30 s), we observe that both sensors choose $b = 0$ with a maximum pulse length. This setting changes the LFM chirp into a purely amplitude modulated signal which has good Doppler estimation properties. The sensor-target geometry prevents the accurate estimation of target range and this configuration permits the tracking algorithm to minimize the Doppler or velocity errors.

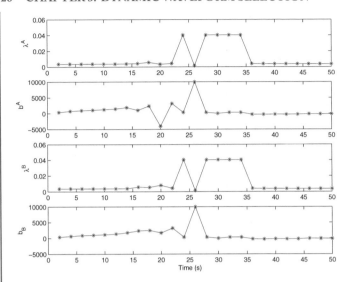

Figure 3.4: Typical time-varying selection of waveform parameters (λ, b) by each sensor for the LFM chirp. Sensor A - top two plots, Sensor B - lower two plots.

Example 2: Agile Phase Functions

In this example, the two sensors independently choose between the three waveforms in Table 3.1 and also dynamically configure their duration and FM rate as in Section 3.5.3. The averaged MSE for this scenario is shown in Figure 3.5, where we compare the performance of the LFM, EFM, and HFM chirps. We observe that the HFM chirp has the best performance of all the waveforms and is accordingly chosen at each tracking instant. In particular, the performance improvement over the LFM is significant.

3.5.4   DISCUSSION

When the waveform is only parameterized by its pulse length, i.e., when the FM rate $b = 0$, the measurement noise covariance matrix becomes diagonal. The measurement errors for position increase while those for velocity decrease with increasing pulse length. This opposing behavior leads to a trade-off between the accuracy of range and range-rate estimation. For the GFM waveforms, however, $b \neq 0$ and the estimation errors in range and range-rate become correlated. When the measurement noise is Gaussian, the conditional variance for range-rate errors given the range may be shown to be

$$\sigma_{\dot{r}|r}^2 = \frac{2c^2}{4\eta f_c^2 \lambda^2}$$

for all pulses, and is only dependent on pulse duration. Using the maximum duration would thus result in the lowest estimation errors for range-rate. The conditional variance on range errors given

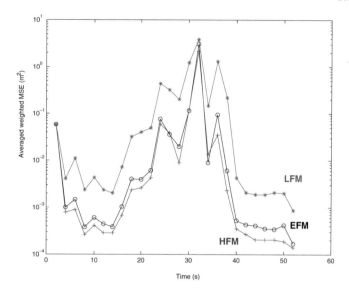

Figure 3.5: Averaged MSE using fixed and configured waveforms and waveform parameters.

the range-rate, however, depends upon $g(\xi)$ in (3.12). It is approximately obtained from $N(\boldsymbol{\theta}_k)$ as

$$
\begin{aligned}
\text{LFM:} \quad \sigma_{r|\dot{r}}^2 &= \frac{c^2}{4\eta\Delta_f^2}\frac{\alpha^2}{2} \\[2mm]
\text{EFM:} \quad \sigma_{r|\dot{r}}^2 &= \frac{c^2}{4\eta\Delta_f^2}\frac{\exp(\lambda(\alpha-\lambda))}{(1+\mathrm{erf}(\lambda))} \\[2mm]
\text{HFM:} \quad \sigma_{r|\dot{r}}^2 &= \frac{c^2}{4\eta\Delta_f^2}\frac{1}{T^2\int_{-\infty}^{\infty}\frac{1}{\lambda\sqrt{\pi}}\exp(-\frac{t^2}{\lambda^2})\frac{1}{(T+|t|)^2}dt}\;.
\end{aligned}
\tag{3.15}
$$

Intuitively, the configuration algorithm must choose the waveform for which $\sigma_{r|\dot{r}}^2$ is the smallest. From Table 3.1, the FM rate for the HFM pulse is not constrained by the value of λ (and thus T_s) in contrast to the LFM and EFM pulses. Thus, it can always be chosen as the maximum allowed value. For a given Δ_f^{max}, T can be chosen large and $\sigma_{r|\dot{r}}^2$ in (3.15) is the lowest for the HFM waveform. It thus provides the best tracking performance as demonstrated in Figure 3.5.

3.6   WIDEBAND ENVIRONMENT

In this section, we consider scenarios where the narrowband condition in (2.14) is not satisfied. This implies that the narrowband model for the received signal developed in Section 2.2.3 is no longer applicable. We introduce the wideband ambiguity function (WBAF) and develop the CRLB based on the WBAF.

3.6.1   WIDEBAND SIGNAL MODEL

When the narrowband condition imposed by (2.14) is not satisfied, the received signal model in (2.15) cannot be used since the scaling of the signal envelope is not negligible. This is especially true in sonar applications where $c \approx 1{,}500$ m/s, and with target velocities in the order of 10 m/s, we must have the time-bandwidth product TBW $\ll 75$ to justify the narrowband assumption. As the time-bandwidth product of sonar signals is generally much higher than 100, wideband processing is necessary. In such situations, the noise-free received signal in (2.13) is given by

$$s_R(t) = \sqrt{2}\mathrm{Re}\left[\sqrt{E_R}\tilde{s}(\sigma t - \tau)\exp\left(j2\pi(f_c\sigma t)\right)\right], \qquad (3.16)$$

where $\sigma = (c - \dot{r})/(c + \dot{r}) > 0$ is the Doppler scale and $\tau = 2r/(c + \dot{r})$ is the delay. Since $\dot{r} \ll c$, we have $r \approx c\tau/2$ and $\dot{r} \approx c(1 - \sigma)/2$.

The maximum difference in amplitude between $\tilde{s}(t(1 - 2\dot{r}/c) - \tau)$ and its approximation $\tilde{s}(t - \tau)$ (when the scaling is neglected) occurs when the difference in their arguments is at its maximum, which is $2\dot{r}\lambda/c$, where λ is the signal duration. If Δ_f^{max} denotes the bandwidth, the signal does not change appreciably in an interval less than $1/\Delta_f^{max}$. Thus, if the signal satisfies $2\dot{r}\lambda/c \ll 1/\Delta_f^{max}$, or $\lambda\Delta_f^{max} \ll c/(2\dot{r})$, the error involved in the approximation is small and the scaling may be neglected. This is the narrowband condition in (2.14). Note, however, that the approximation $\sigma \approx 1$ cannot be made in the phase in (3.16). Accordingly, substituting $\sigma = 1 - (2\dot{r}/c)$ in (3.16), and ignoring the scaling, we obtain the narrowband version of the received signal in (2.15). Thus, (2.15) is an approximation of (3.16).

When the environment causes wideband changes in the signal, we must use a processing tool to match it. In this case, the wideband ambiguity function provides a description of the correlation properties of the signal, and it is given by [48]

$$WAF_{\tilde{s}}(\tau, \sigma) = \sqrt{\sigma}\int_{-\infty}^{\infty}\tilde{s}(t)\tilde{s}^*(\sigma t + \tau)dt. \qquad (3.17)$$

As in the narrowband case, the Fisher information matrix is obtained by taking the negative of the second derivatives of the WBAF, evaluated at the true target delay and scale [49, 50]. Alternatively, the elements of the Fisher information matrix can be derived directly from the waveform as shown in Appendix B.

3.6.2   WAVEFORM STRUCTURE

Upon closer analysis of the results presented in Section 3.5.3, we found that the amplitude envelope of the waveform plays a significant role in its performance from a tracking perspective. A Gaussian envelope has often been used in the past [17, 18, 19, 21, 47, 23]   due to the flexibility it affords in computing ambiguity functions and moments of the time and frequency distributions of the signal. However, the behavior of the phase function in the tails of the Gaussian envelope does not significantly affect the waveform's performance in range and Doppler estimation, since the tails contain only a small fraction of the signal energy. This can lead to biased conclusions on waveform performance when different phase functions are considered. In Table 3.1, the LFM and EFM chirps have larger frequency deviation towards the tails of the envelope while the HFM chirp causes larger frequency deviations towards the center of the envelope. This explains the superior performance of the HFM chirp in Section 3.5.3. In order to have a fairer comparison between different phase functions, it is necessary to use an amplitude envelope that is uniform. Accordingly, a rectangular envelope would be most desirable, but the CRLB associated with such waveforms cannot be computed in closed form since the second moment of its spectrum is not finite [51].

In this and the next chapter, we use a trapezoidal envelope because it avoids the difficulties associated with the evaluation of the CRLB for waveforms with a rectangular envelope and yet approximates it well enough to provide a clearer comparison of phase function performance than using a Gaussian envelope. The complex envelope of the signal is defined by (2.12) with

$$
a(t) = \begin{cases} \frac{\alpha}{t_f}(\frac{T_s}{2} + t_f + t), & -T_s/2 - t_f \leq t < -T_s/2 \\ \alpha, & -T_s/2 \leq t < T_s/2 \\ \frac{\alpha}{t_f}(\frac{T_s}{2} + t_f - t), & T_s/2 \leq t < T_s/2 + t_f, \end{cases} \tag{3.18}
$$

where α is an amplitude chosen so that $\tilde{s}(t)$ in (2.12) has unit energy. The finite rise/fall time of $a(t)$ is $t_f \ll T_s/2$. Note that the trapezoidal envelope in (3.18) closely approximates a rectangular pulse but the finite rise/fall time t_f permits the evaluation of the CRLB. We define the chirp duration as $\lambda = T_s + 2t_f$. Note that the frequency sweep Δ_f can be specified by b, λ and $\xi(t)$. In order to completely specify a time-frequency signature for the waveform, we must also fix either the initial frequency $f(-\lambda/2) = f_1$ or the final frequency $f(\lambda/2) = f_2$. An additional parameter γ is therefore introduced in the phase functions of the considered waveforms which are shown in Table 3.2. For example, in the case of the LFM, $f(-\lambda/2) = b/\gamma = f_1$ and $f(\lambda/2) = f_1 + b\lambda = f_2$ can be used to uniquely determine b and γ once λ, Δ_f and either f_1 or f_2 are chosen. We fix the larger of f_1 and f_2 and vary the remaining parameters to obtain various time-frequency characteristics.

Table 3.2: GFM waveforms used in the configuration scenarios. The exponent κ defines the time-frequency relationship of the power-law FM (PFM) chirp waveform.

Waveform	Phase Function $\xi(t)$	Frequency sweep Δ_f
LFM	$t/\gamma + (t + \lambda/2)^2/2$	$b\lambda$
PFM	$t/\gamma + (t + \lambda/2)^\kappa/\kappa$	$b\lambda^{\kappa-1}$
HFM	$\ln(t + \gamma + \lambda/2)$	$b\frac{\lambda}{\gamma(\gamma+\lambda)}$
EFM	$\exp(-(t + \lambda/2)/\gamma)$	$b\frac{1}{\gamma}(\exp(-\frac{\lambda}{\gamma}) - 1)$

3.6.3   SIMULATION

The simulation study [52] to test the performance of the waveform scheduling algorithm in a wideband environment involved two waveform-agile sensors tracking an underwater target moving in two dimensions, using range and range-rate measurements. In the first example, the library consists of the HFM chirp only, while in the second example, the selection of the waveform phase function is also permitted.

The carrier frequency of the transmitted waveform was $f_c = 25$ kHz and the waveform duration ranged within 0.01 s $\leq \lambda \leq 0.3$ s. The SNR at sensor i at time k was determined according to (3.14), where $r_0 = 500$ m was the range at which 0 dB SNR was obtained. We considered a number of GFM chirps with different time-frequency signatures that are specified by their phase functions, $\xi(t/t_r)$ in Table 3.2 with $t_r = 1$. We define the frequency sweep to be $\Delta^i_f = |v^i(\lambda^i/2) - v^i(-\lambda^i/2)|$, where $v^i(t)$ is the instantaneous frequency, and limit it to a maximum of 2 kHz. We fix $v^i(-\lambda^i/2) = f_c + 2$ kHz (downswept chirps) or $v^i(\lambda^i/2) = f_c + 2$ kHz (upswept chirps). We then compute b^i and γ^i in Table 3.2 for each waveform for any chosen frequency sweep. The sampling interval and the process noise intensity in (2.10) were $\Delta T_s = 2$ s and $q = 0.01$, respectively. The speed of sound in water was taken as $c = 1,500$ m/s. The UPF was used as the target tracker and the weighting matrix in (2.19) was set to $\Lambda = \text{diag}[1, 1, 4 \text{ s}^2, 4 \text{ s}^2]$ so that the cost was in units of m^2. All results were averaged over 500 runs. The trajectory of the target is shown in Figure 3.6.

Example 1: HFM Only

In the first example, we configured each sensor with the HFM waveform and dynamically selected its parameters (λ and b) for each sensor so as to minimize the predicted MSE. The actual MSE obtained in tracking a target, averaged over 500 simulations, is shown in Figure 3.7. For comparison, we also show the MSE that was obtained when the sensor configurations were fixed at the minimum or maximum durations and the maximum frequency sweep. The improvement in performance in using dynamic parameter selection is apparent. Figure 3.8 shows the dynamic selection of the frequency sweep that results from the selection of the chirp rate b^i by the configuration algorithm. Note that the frequency sweep chosen by dynamically selecting the chirp rate is much less than the maximum of 2 kHz. The pulse duration was always selected as the maximum of 0.3 s.

Example 2: Agile Phase Functions

In the second simulation example, we permitted the sensors to dynamically select the phase function from among the options in Table 3.2. We allowed the waveform selection to choose between hyperbolic, exponential, and power-law FM chirps with $\kappa = 2$ and 2.6, in addition to choosing the duration λ and chirp rate b so as to minimize the predicted MSE. Figure 3.9 plots the averaged MSE when dynamic parameter selection is used with FM waveforms with different phase functions. Note that the power-law FM (PFM) chirp with $\kappa = 2$ constitutes a linear FM chirp. From Figure 3.9, the PFM chirp with $\kappa = 2.6$ performs the best, and we should expect that it will always be chosen as it was indeed observed. The HFM chirp yielded the poorest performance as seen in Figure 3.9. It has been shown that the HFM chirp is optimally Doppler tolerant [35]. This implies that it is minimally Doppler sensitive and should compare poorly with the tracking performance of other FM waveforms.

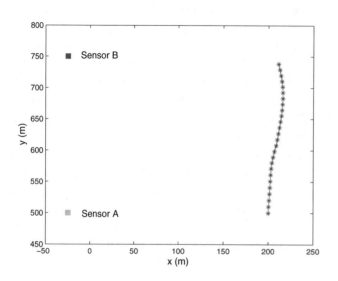

Figure 3.6: Target trajectory with Sensor A and B located at (-25,500) m and (-25,750) m, respectively.

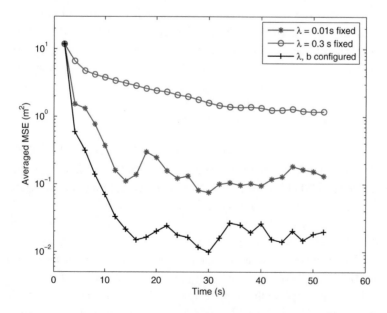

Figure 3.7: Averaged MSE when both sensors use HFM chirps.

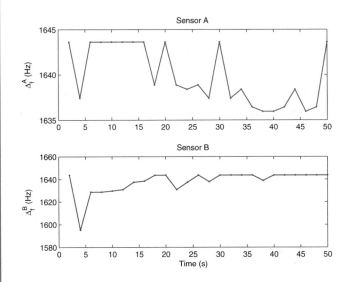

Figure 3.8: Dynamic selection of the frequency sweep Δ_f^i, when both sensors $i = $ A, B use HFM waveforms [52].

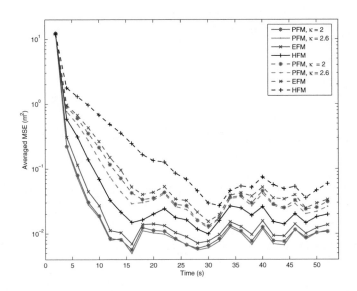

Figure 3.9: Comparison of averaged MSE using various waveforms when the allowed frequency sweep is 1 kHz (dotted) and 2 kHz (solid) [52].

3.6.4   DISCUSSION

As the variance of the range-rate estimation errors depends inversely on the pulse duration, the algorithm attempts to minimize these errors when the maximum allowed pulse length is used. On the other hand, range estimation errors can be minimized by using the maximum time-bandwidth product. However, the correlation between the errors increases with increasing frequency sweep, thereby reducing the ability of the waveform to estimate range and range-rate *simultaneously*. Thus, there is a trade-off in the choice of frequency sweep and this is reflected in the selections made by the configuration algorithm.

CHAPTER 4

Dynamic Waveform Selection for Tracking in Clutter

The tracking of a target with observations that do not guarantee perfect detection in an environment that includes clutter is significantly complicated by the uncertainty in the origin of the measurements. This introduces new trade-offs in the waveform selection process [53]. In this chapter, we examine the problem of selecting and configuring waveforms for tracking targets in clutter with imperfect detections. We present a methodology to predict the mean square error (MSE) in this scenario. Using the probabilistic data association filter, the algorithm is first applied to the tracking of a single target and then extended to multiple targets.

4.1 SINGLE TARGET

In this section, we present the application of the waveform-agile tracking algorithm in an environment that includes a single target that is observed in the presence of clutter with nonunity probability of detection. We begin by augmenting the observation model described in Section 2.2.3 to reflect the fact that at each sampling instant, each sensor obtains more than one measurement. We also introduce a model for false alarms due to clutter.

4.1.1 OBSERVATIONS MODEL

When clutter is present in the observation space, there are likely to be false alarms due to clutter. Since we also assume imperfect detection, sensor i will obtain $m_k^i \geq 0$ measurements at time k. We define the measurement vector for sensor i at time k as

$$\mathbf{Z}_k^i = [\mathbf{z}_{k,i}^1, \ \mathbf{z}_{k,i}^2, \ \ldots, \mathbf{z}_{k,i}^{m_k^i}] \,, \tag{4.1}$$

where $\mathbf{z}_{k,i}^m = [r_k^{i,m} \ \dot{r}_k^{i,m}]^T, m = 1, \ldots, m_k^i$. We also define $\mathbf{Z}_k = [\mathbf{Z}_k^A \ \mathbf{Z}_k^B]$ as the full measurement at time k. The probability that the target is detected (i.e., that $\mathbf{Z}_k^i$ includes a target-originated measurement) is given by $P_{d_k}^i$. The target-originated measurements at each sensor are given by (2.16).

4.1.2 CLUTTER MODEL

We assume that the number of false alarms follows a Poisson distribution with average ρV_k^i, where ρ is the density of the clutter and V_k^i is the validation gate volume. The validation gate is a region in the observation space centered on the next predicted observation [3]. Only observations that fall within this region are validated as having been potentially originated by the target. The probability that m false alarms are obtained is

$$\mu(m) = \frac{\exp(-\rho V_k^i)(\rho V_k^i)^m}{m!} .$$ (4.2)

We also assume that the clutter is uniformly distributed in the volume V_k^i. The probability of detection at time k is modeled according to the relation

$$P_{d_k}^i = P_f^{\frac{1}{1+\eta_k^i}} ,$$

where P_f is the desired probability of false alarm and η_k^i is the SNR at sensor i.

4.1.3   TARGET TRACKING WITH PROBABILISTIC DATA ASSOCIATION

As described in Section 2.1.1, we use a particle filter as the tracker. As is often the case, we use the kinematic prior, $p(\mathbf{X}_k^j|\mathbf{X}_{k-1}^j)$, as the importance density. From (2.5), the weights are then proportional to the likelihood function $p(\mathbf{Z}_k|\mathbf{X}_k^j, \boldsymbol{\theta}_k)$ as in (2.7). Next, we present the likelihood function for this scenario.

Since there is uncertainty in the origin of the measurements, the target tracker must also estimate the measurement-to-target association. One method of filtering in this scenario is to treat each individual measurement, in turn, as target-originated, and all the other measurements as being due to clutter. This gives rise to m_k^i associations, each of which is weighted with its probability given the measurements, and incorporated in the filter update. This approach is known as probabilistic data association [3] and computes the likelihood as an average over all possible data associations. The measurements obtained at each sensor are assumed to be independent. Therefore, the likelihood function is the product of the likelihood functions for each sensor and is given by

$$p(\mathbf{Z}_k|\mathbf{X}_k, \boldsymbol{\theta}_k) = \prod_{i=A,B} p(\mathbf{Z}_k^i|\mathbf{X}_k, \boldsymbol{\theta}_k^i) .$$ (4.3)

To derive $p(\mathbf{Z}_k^i|\mathbf{X}_k, \boldsymbol{\theta}_k^i)$, we assume that the target generates at most one measurement which is detected with probability $P_{d_k}^i$, while the remaining measurements consist of false alarms due to clutter. The measurement $\mathbf{Z}_k^i$ must therefore be comprised of either one target-originated measurement together with $m_k^i - 1$ false alarms, or m_k^i false alarms. In the former case, the probability that any one of the m_k^i measurements was generated by the target is equal, and as a result, all measurement-to-target associations are equiprobable. The probability of obtaining the total measurement $\mathbf{Z}_k^i$, given that the target was detected, will therefore include the probability that each individual measurement $\mathbf{z}_{k,i}^m, m = 1, \ldots, m_k^i$ in (4.1) was generated by the target, weighted by the measurement's association probability. Thus, we can directly write the likelihood function for sensor i in (4.3) as [54]

$$
\begin{aligned}
p(\mathbf{Z}_k^i | \mathbf{X}_k, \boldsymbol{\theta}_k^i) \;=\;& (1 - P_{d_k}^i)\mu(m_k^i)(V_k^i)^{-m_k^i} \\[4pt]
& + P_{d_k}^i (V_k^i)^{-(m_k^i-1)}\mu(m_k^i - 1) \cdot \frac{1}{m_k^i} \sum_{m=1}^{m_k^i} p(\mathbf{z}_{k,i}^m | \mathbf{X}_k, \boldsymbol{\theta}_k^i) \, .
\end{aligned}
\tag{4.4}
$$

4.1.4   WAVEFORM SELECTION IN THE PRESENCE OF CLUTTER

The algorithm for waveform selection presented in Chapter 3 assumes that there is only one measurement and that it is target-originated. While the algorithm structure remains the same when clutter and imperfect detection are considered, the computation of the predicted cost must reflect the uncertainty in the origin of the measurements. Next, we describe the method used to approximate the predicted cost using the unscented transform.

We assume that the sigma points $\boldsymbol{\mathcal{X}}_n$ corresponding to a distribution with mean $\hat{\mathbf{X}}_{k|k-1}$ and covariance $P_{k|k-1}$ have been computed according to (3.7). As before, we wish to approximate the covariance $P_{k|k}(\boldsymbol{\theta}_k)$ that would be obtained if a waveform characterized by its parameter vector $\boldsymbol{\theta}_k$ is used to obtain a measurement at time k. For sensor i, a transformed set of sigma points $\boldsymbol{\mathcal{Z}}_n^i = h^i(\boldsymbol{\mathcal{X}}_n)$ is computed. Then, as in (3.8) and (3.9), we calculate the covariances

$$
P_{zz}^i \;=\; \sum_{n=0}^{2N+1} \mathcal{W}_n \left(\boldsymbol{\mathcal{Z}}_n^i - \bar{\boldsymbol{\mathcal{Z}}} \right)\left(\boldsymbol{\mathcal{Z}}_n^i - \bar{\boldsymbol{\mathcal{Z}}} \right)^T ,
\tag{4.5}
$$

$$
P_{xz}^i \;=\; \sum_{n=0}^{2N+1} \mathcal{W}_n \left(\boldsymbol{\mathcal{X}}_n - \bar{\boldsymbol{\mathcal{X}}} \right)\left(\boldsymbol{\mathcal{Z}}_n^i - \bar{\boldsymbol{\mathcal{Z}}} \right)^T ,
\tag{4.6}
$$

where

$$
\bar{\boldsymbol{\mathcal{Z}}} = \sum_{n=0}^{2N+1} \mathcal{W}_n \boldsymbol{\mathcal{Z}}_n^i \quad \text{and} \quad \bar{\boldsymbol{\mathcal{X}}} = \sum_{n=0}^{2N+1} \mathcal{W}_n \boldsymbol{\mathcal{X}}_n \, .
$$

Note that the expectation in (2.19) is also over all realizations of the clutter and thus over all possible measurement-to-target associations. The calculation of $P_{k|k}(\boldsymbol{\theta}_k)$ should reflect this fact. Intuitively, if there was no uncertainty in the origin of the measurements, all the information contained in the target-originated measurement could be unambiguously used to reduce the covariance of the state estimate as in (3.10). As is evident from (4.4), each measurement, whether target-generated or not, contributes to the update. The measurements that are not target-generated thus limit the reduction in the covariance during the update step. When probabilistic data association is used, it has been shown that the expected covariance of the state after an update with the measurement at sensor i is given by [55]

$$
P_{k|k}^i(\boldsymbol{\theta}_k^i) = P_{k|k-1} - q_{2_k}^i P_{k|k}^c \, ,
\tag{4.7}
$$

where $P^c_{k|k}$ is the update due to the true measurement. Using the unscented transform, this update is given by

$$P^c_{k|k} = P^i_{xz}(P^i_{zz} + (N(\theta^i_k))^{-1} P^{i\,T}_{xz} .$$

(4.8)

The scalar $q^i_{2_k}$ in (4.7) lies between 0 and 1 and is called the information reduction factor. From (3.10), (4.7), and (4.8), we see that it serves to lower the reduction in the covariance that would have been obtained if there was no uncertainty in the origin of the measurements due to clutter. The information reduction factor depends upon $P^i_{d_k}$, ρ and V^i_k, and its computation involves a complicated integral which has to be evaluated by Monte Carlo methods. However, in this work, we use some approximations that are available in the literature [56].

Note that the information reduction factor is different for each sensor. This requires that the update in (4.7) be performed sequentially for each sensor. Accordingly, we first use the measurement of Sensor A to obtain $P^A_{k|k}(\theta^A_k)$ using (4.7) and (4.8). A second update, using the measurements of Sensor B, is then carried out with $P_{k|k-1} = P^A_{k|k}(\theta^A_k)$ in (4.7) to yield $P^B_{k|k}(\theta^B_k)$. The second update requires another computation of all the covariance matrices in (4.8) using the unscented transform with $i = B$. Note that $P^B_{k|k}(\theta^B_k)$ is the final predicted covariance $P_{k|k}(\theta_k)$ as it includes information from both sensors A and B. The approximate cost $J(\theta_k)$ in (2.19) is then computed as the trace of $\Lambda P_{k|k}(\theta_k)$.

4.1.5   SIMULATION

The waveform library used in this simulation is the same as that used in Section 3.6.3, and the phase functions and the bandwidth are shown in Table 3.2. In this example, however, we use radar sensors and assume that the received signal satisfies the narrowband assumption. Figure 4.1 shows the instantaneous frequencies of these waveforms for a duration of $\lambda = T_s + 2t_f = 100\ \mu s$ and a frequency sweep of $\Delta_f = 14$ MHz.

Our simulations consist of two radar sensors tracking a target that moves in two dimensions. The signal duration was restricted to the range $10\ \mu s \leq \lambda^i_k \leq 100\ \mu s$ while the carrier frequency was $f_c = 10.4$ GHz. For each waveform considered, the frequency sweep was determined by selecting the start and end points of the desired time-frequency signature as in Figure 4.1, with the restriction that the maximum frequency sweep was 15 MHz. The FM rate b^i_k and the parameter γ^i in Table 3.2 were then computed from the frequency sweep and the duration of the waveform. The sampling interval in (2.10) was $\Delta T_s = 250$ ms. The SNR was computed according to (3.14) where $r_0 = 50$ km was the range at which a 0 dB SNR was obtained and the probability of false alarm was $P_f = 0.01$. The validation gate was taken to be the 5-sigma region around the predicted observation [3]. The target trajectory was generated with the initial state $X_0 = [0\ 0\ 100\ 800]^T$. The process noise intensity in (2.10) was $q = 0.1$, while the covariance of the initial estimate provided to the tracker was $P_0 = \mathrm{diag}[1,000\ 1,000\ 50\ 50]$. Sensor A was located at $(0, -15,000)$ m and Sensor B at $(15,631\ 4,995)$ m. The standard particle filter was used as the tracker, and the weighting matrix in (2.19) was set to $\Lambda = \mathrm{diag}[1\ 1\ 1\,s^2\ 1\,s^2]$ so that the cost function in (2.19) had units of m^2. All results were

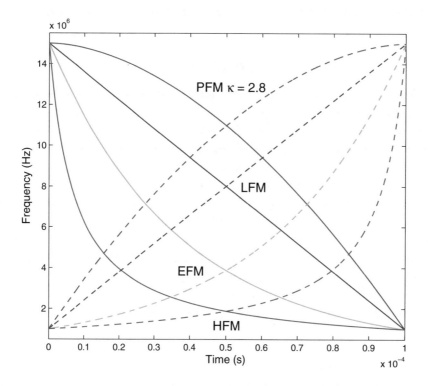

Figure 4.1: Time-frequency plots of generalized FM chirp waveforms (downsweep - solid lines, upsweep - dotted lines) with duration 100 μs and frequency sweep 14 MHz.

averaged over 500 Monte Carlo simulations. When more than four continuous measurements fell outside the validation gate for either sensor, the track was categorized as lost.

Example 1: Linear Frequency Modulated (LFM) Chirp Only
In the first example, we consider the performance of the dynamic parameter configuration algorithm when only the LFM chirp is transmitted by the sensors. Specifically, we compare the performance of configuring the duration and FM rate (or equivalently frequency sweep) of the transmitted LFM chirp to that obtained when an LFM chirp is used with fixed parameters. For the fixed LFM chirps, we consider the shortest and longest allowable durations and the maximum allowed frequency sweep. The averaged MSE, computed from the actual error in the estimate and conditioned on convergence due to possible lost tracks, is shown in Figure 4.2 for two clutter densities, $\rho = 0.0001$ and $\rho = 0.001$ false alarms per unit validation gate volume. It is apparent that the configuration algorithm provides improved performance. lost by the tracker. Specifically, when the fixed waveforms were used, an average of 25% and 38% of tracks were lost by the 10 μs and 100 μs pulse, respectively. The dynamically configured waveform, however, lost only 0.7% of the tracks. This can be attributed to

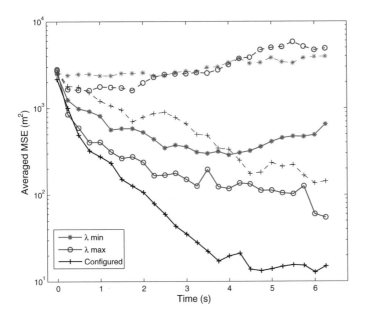

Figure 4.2: Comparison of averaged MSE using fixed and configured LFM chirps only. Clutter density $\rho = 0.0001$ (solid lines) and $\rho = 0.001$ (dotted lines) [22].

the number of false alarms that appear in the measurements at each sensor, which in turn is directly proportional to the validation gate volume. In Figure 4.3 we compare the volume for the three LFM waveforms for $\rho = 0.0001$, and find that the configured waveform results in a much lower volume for both sensors. The validation gate volume for sensor i is proportional to the determinant of the matrix $(P_{zz}^i + N(\boldsymbol{\theta}_k^i))$ in (4.8). As the search over the space of waveforms proceeds, it is apparent that this determinant will be minimized. In addition, the information reduction factor q_2^i in (4.7) increases as the validation gate volume decreases [55]. Thus, a waveform that results in a smaller validation gate also contributes more towards reducing the covariance of the state estimate.

Example 2: Agile Phase Function
In this simulation, in addition to dynamically configuring the duration and FM rate, we also select the phase function $\xi(t/t_r)$ in (2.12) of the generalized chirp waveforms. The set of waveforms available to the sensors includes the power-law FM (PFM) chirp with varying κ, the hyperbolic FM (HFM) and exponential FM (EFM) chirps as described in Table 3.2. For the simulations, we chose $\kappa = 2, 2.8, 3$. Recall that $\kappa = 2$ yields an LFM waveform. For this example the clutter density was $\rho = 0.0001$ false alarms per unit validation gate volume. Figure 4.4 shows the averaged MSE, conditioned on convergence, that is obtained when the phase function is dynamically selected, in addition to the waveform duration and FM rate. For comparison, the averaged MSE for each

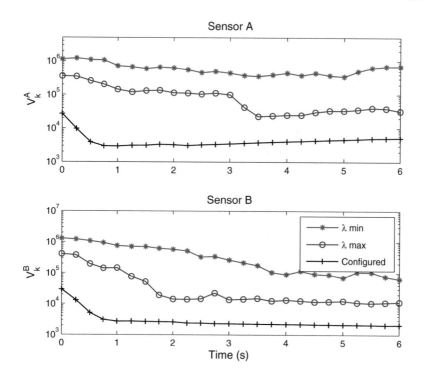

Figure 4.3: Comparison of averaged validation gate volume using fixed and configured LFM chirps only with clutter density $\rho = 0.0001$ [22].

generalized chirp is also shown; this is obtained when the generalized FM chirp has its duration and FM rate configured as in Example 1 for the LFM.

We find that there is some improvement in the tracking performance when the phase function is dynamically selected. A typical waveform selection for one simulation run is shown in Figure 4.5. We note that the HFM and the EFM chirps are chosen at the first few instants while the PFM chirp with $\kappa = 3$, and hence parabolic instantaneous frequency, is used later. The LFM chirp and the PFM chirp with $\kappa = 2.8$ are not selected at all.

4.1.6 DISCUSSION

In order to understand the waveform selection behavior, note that when the tracking starts, the range and range-rate of the target are poorly known and must be simultaneously estimated. Waveforms that offer little correlation between the estimation errors should accordingly be selected. Figure 4.6 shows the correlation coefficient for the waveforms available to the tracker for a duration of 100 μs and a frequency sweep of 15 MHz. Note that the HFM and EFM chirps have smaller correlation coefficients than the PFM chirps. When the range and range-rate estimates of the targets

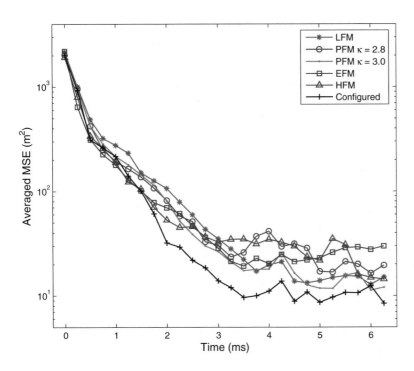

Figure 4.4: Comparison of averaged MSE, conditioned on convergence, with and without phase function agility [22].

improve, it is possible to utilize the correlation between the estimation errors to improve the tracking performance. At this stage, waveforms that provide considerable correlation become more suitable. Thus, the conditional variance of the range given the range-rate becomes an important feature. As seen in Figure 4.6, the PFM chirps offer lower values than the EFM and HFM chirps. However, minimization of the validation gate volume also occurs during the dynamic selection procedure. The volume for $P_{zz} = \mathrm{diag}[5, -50, -50, 1,000]$ and an SNR of 20 dB is also shown in Figure 4.6. Note that the PFM chirp with $\kappa = 3$ offers a lower volume than the LFM chirp and is therefore always selected during the later stages of the tracking sequence.

4.1.7 PERFORMANCE UNDER DIFFERENT CHOICES OF THE WEIGHTING MATRIX

The choice of the weighting matrix Λ in (2.19) can have significant impact on the performance of the algorithm. In order to investigate this dependence, we conducted simulations in which the

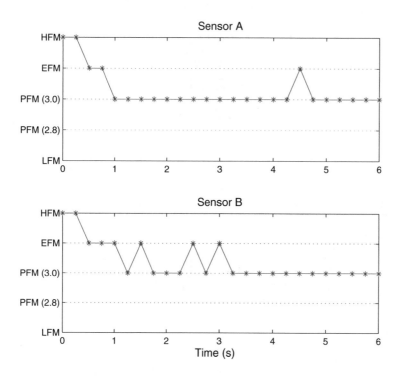

Figure 4.5: Typical waveform selection when the phase function is configured together with the duration and FM rate of the waveform [22].

cost function depended only on position variance ($\Lambda = \mathrm{diag}[1, 1, 0, 0]$), and only on velocity variance ($\Lambda = \mathrm{diag}[0, 0, 1, 1]$).

Figure 4.7 compares the position MSE achieved by the phase-function-agile waveform under different choices of Λ in (2.19). In Figure 4.7, note that the performance is best when $\Lambda = \mathrm{diag}[1, 1, 1\,\mathrm{s}^2, 1\,\mathrm{s}^2]$, i.e., when both position and velocity variances contribute to the cost function. This is due to the fact that the target tracker uses both velocity and position estimates at time $k - 1$ to estimate the position at time k. A poor estimate of velocity thus also contributes to position errors. The performance in velocity estimation in various cases is shown in Figure 4.8. It can be noted from Figure 4.7 that the difference in performance is the greatest between approximately 1.75 s and 4.75 s during the tracking sequence. This is due to the fact that during this period, the target is furthest from both sensors leading to poor SNR. Thus, the estimation errors in range and range-rate are high. Accordingly, the tracker places a greater dependence on the kinematics model, which, as stated earlier, can only provide accurate information if both position and velocity are well estimated. A typical waveform selection sequence when the cost function depends on position variance only and velocity variance only is shown in Figures 4.9 and 4.10, respectively. It is interesting

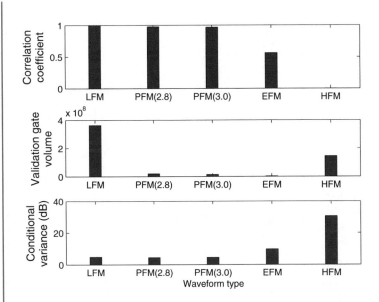

Figure 4.6: Correlation coefficient (top), validation gate volume (middle), and conditional variance of range given range-rate (bottom) for the waveforms used in Section 4.1.5. The correlation coefficient for the HFM chirp is negligibly small [22].

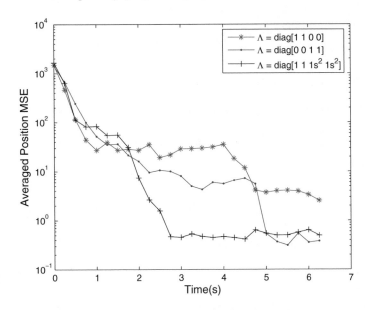

Figure 4.7: Comparison of averaged position MSE, conditioned on convergence, under different choices of the weighting matrix $\boldsymbol{\Lambda}$ in (2.19).

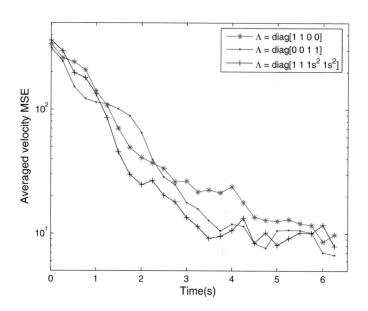

Figure 4.8: Comparison of averaged velocity MSE, conditioned on convergence, under different choices of the weighting matrix Λ in (2.19).

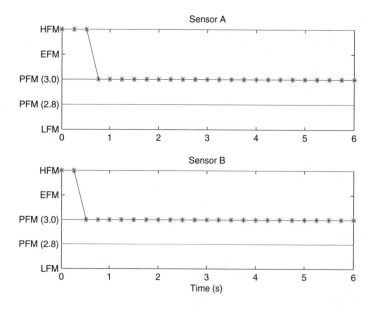

Figure 4.9: Typical waveform selection when the cost function depends on position variance alone.

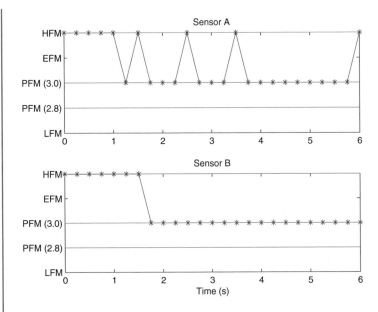

Figure 4.10: Typical waveform selection when the cost function depends on velocity variance alone.

to note that the behavior is similar to that observed in Figure 4.5 for the case where both variances contribute to the cost function.

4.2   MULTIPLE TARGETS

In this section, we extend the waveform selection algorithm developed in Section 4.1 to a scenario that includes multiple targets, clutter and missed detections [57]. Specifically, assuming that the number of targets is known, we show how waveform selection can be used to minimize the total MSE of tracking multiple targets and present a simulation study of its application to the tracking of two targets. We begin by extending the target dynamics and observations models of Sections 2.2.1 and 4.1.1, respectively, to multiple targets.

4.2.1   TARGET DYNAMICS
Let

$$\mathbf{X}_k = [\mathbf{x}_k^{1^T}, \ldots, \mathbf{x}_k^{S^T}]^T \tag{4.9}$$

represent the state of S targets that move in a two-dimensional space. The state of target s, $s = 1, \ldots, S$ is $\mathbf{x}_k^s = [x_k^s \ y_k^s \ \dot{x}_k^s \ \dot{y}_k^s]^T$, where x_k^s and y_k^s correspond to the position, and $\dot{x}_k^s$ and $\dot{y}_k^s$ to the velocity at time k in Cartesian coordinates. The dynamics of each target are modeled by a linear, constant velocity model given by

$$\mathbf{x}_k^s = F \mathbf{x}_{k-1}^s + \mathbf{w}_k^s \, , \tag{4.10}$$

where $\mathbf{w}_k^s$ is a white Gaussian noise sequence that models the process noise. The constant matrix F and the process noise covariance matrix Q are as given by (2.10).

4.2.2   MEASUREMENT MODEL

At each sampling epoch, both sensors, Sensor A and Sensor B, obtain range (r) and range-rate ($\dot{r}$) measurements of each target. At time k, the measurement corresponding to target s obtained at sensor i, is given by

$$\mathbf{z}_k^{s,i} = h^i(\mathbf{x}_k^s) + \mathbf{v}_k^{s,i} \, , \tag{4.11}$$

where $\mathbf{v}_k^{s,i}$ is a white Gaussian noise sequence that models the measurement error. We assume that $\mathbf{v}_k^{s,i}$ in (4.11) is independent between all targets and sensors, and that $\mathbf{v}_k^{s,i}$ and $\mathbf{w}_k^s$ in (4.10) are uncorrelated for all targets. The nonlinear function $h^i(\mathbf{x}_k^s) = [r_k^{s,i} \ \dot{r}_k^{s,i}]^T$, where

$$\begin{aligned}
r_k^{s,i} &= \sqrt{(x_k^s - x^i)^2 + (y_k^s - y^i)^2} \\
\dot{r}_k^{s,i} &= (\dot{x}_k^s(x_k^s - x^i) + \dot{y}_k^s(y_k^s - y^i))/r_k^{s,i} \, ,
\end{aligned}$$

and sensor i is located at (x^i, y^i).

4.2.3   CLUTTER MODEL

As in Section 4.1.2, we assume that the number of false alarms at sensor i follows a Poisson distribution with average ρV_k^i, where ρ is the density of the clutter, and V_k^i is the validation gate volume [3]. The validation gate volume for target s at sensor i is $V_k^{s,i}$, and V_k^i is the union of the volume for each of the S targets at sensor i. The probability that m false alarms are obtained is given by (4.2). We also assume that the clutter is uniformly distributed in the volume V_k^i. The probability of detection at time k is modeled according to the relation

$$P_{d_k}^{s,i} = P_f^{\frac{1}{1+\eta_k^{s,i}}} \, , \tag{4.12}$$

where P_f is the desired false alarm probability and $\eta_k^{s,i}$ is the SNR at sensor i corresponding to target s.

4.2.4   MULTIPLE TARGET TRACKING WITH JOINT PROBABILISTIC DATA ASSOCIATION

We use a particle filter tracker to recursively estimate the probability distribution $p(\mathbf{X}_k | \mathbf{Z}_{1:k}, \boldsymbol{\theta}_{1:k})$ of the target state, given the sequence of observations $\mathbf{Z}_{1:k}$ and waveform vectors $\boldsymbol{\theta}_{1:k}$ up to time k. Each particle consists of a concatenation of the states of each individual target and is thus a

$4S \times 1$ vector. When multiple targets are to be tracked in the presence of clutter, the measurement-to-target association must also be estimated. Using arguments similar to Section 4.1.3 for a single target in clutter, we present the likelihood function for the case of two targets in clutter. With the kinematic prior used as the sampling density as in (2.6), the weights are proportional to the likelihood function. Since the measurement noise at each sensor is assumed to be independent, the total likelihood function is given by (4.3), and we need only derive $p(\mathbf{Z}_k^i|\mathbf{X}_k, \boldsymbol{\theta}_k^i)$, the likelihood function for sensor i.

When the targets are well separated in the observation space, the likelihood function factorizes into S components, one for each target, so that

$$p(\mathbf{Z}_k^i|\mathbf{X}_k, \boldsymbol{\theta}_k^i) = \prod_{s=1}^{S} p(\mathbf{Z}_k^{s,i}|\mathbf{x}_k^s, \boldsymbol{\theta}_k^i) \,, \tag{4.13}$$

where $\mathbf{Z}_k^{s,i} \subset \mathbf{Z}_k^i$ is a set consisting of the observations in the validation gate of target s. Such a scenario corresponds to the case where the validation gates for each of the targets do not intersect. In this situation, the joint tracking of the targets reduces to tracking each target independently [58]. Assuming that each target produces at most one measurement, the likelihood functions in (4.13) are as in (4.4)

$$\begin{aligned} p(\mathbf{Z}_k^{s,i}|\mathbf{x}_k^s, \boldsymbol{\theta}_k^i) &= (1 - P_k^{s,i})\mu(n_s)(V_k^{s,i})^{-n_s} \\ &+ P_k^{s,i}(V_k^{s,i})^{-(n_s-1)} \cdot \mu(n_s - 1)\frac{1}{n_s}\sum_{m=1}^{n_s} p_m^{s,i} \,, \end{aligned}$$

where n_s is the number of observations in the validation gate for target s and $p_m^{s,i} = p(\mathbf{z}_{k,m}^{s,i}|\mathbf{x}_k^s, \boldsymbol{\theta}_k^i)$ is the probability that the mth measurement in $\mathbf{Z}_k^{s,i}$ (or equivalently, the validation gate region for target s) was originated from target s.

When the targets are not well separated, their validation gates may overlap, as shown in Figure 4.11 for $S = 2$ targets. Here, n_1 and n_2 are the number of measurements that are validated exclusively for Target 1 and Target 2, respectively, while n_3 is the number of measurements validated for both targets. In this case, the factorization in (4.13) does not hold, and the tracker must estimate the states of each target jointly. We then have

$$p(\mathbf{Z}_k^i|\mathbf{X}_k, \boldsymbol{\theta}_k^i) = \sum_{r=1}^{R} p(\mathbf{Z}_k^i|\Omega_r, \mathbf{X}_k, \boldsymbol{\theta}_k^i)p(\Omega_r) \,,$$

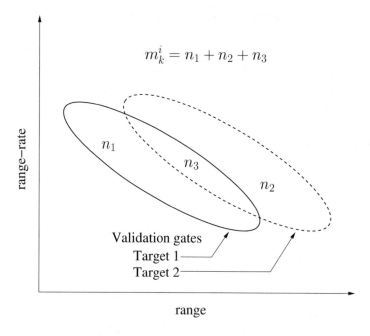

Figure 4.11: Overlapping validation gates in observation space for two targets.

where Ω_r is the measurement-to-target association [3]. By exhaustively enumerating all possible associations when both targets, either target, or no targets are detected, it can be shown that

$$
\begin{aligned}
p(\mathbf{Z}_k^i | \mathbf{X}_k, \boldsymbol{\theta}_k^i) \quad \propto \quad & (1 - P_{d_k}^{1,i})(1 - P_{d_k}^{2,i}) \\
+ \quad & \frac{m_k^i \rho^{-1} P_k^{1,i} (1 - P_k^{2,i})}{n_1 + n_3} \sum_{m=1}^{n_1+n_3} p_m^{1,i} \\
+ \quad & \frac{m_k^i \rho^{-1} P_k^{2,i} (1 - P_k^{1,i})}{n_2 + n_3} \sum_{m=1}^{n_2+n_3} p_m^{2,i} \\
+ \quad & \frac{m_k^i (m_k^i - 1) \rho^{-2} P_k^{1,i} P_k^{2,i}}{n_1 n_2 + n_2 n_3 + n_1 n_3} \sum_{l=1}^{n_1+n_3} \sum_{m=1}^{n_2+n_3} p_l^{1,i} p_m^{2,i} .
\end{aligned}
\tag{4.14}
$$

Note that each target produces at most one measurement and the last summation in (4.14) must exclude all combinations that violate this fact. Although this was only shown for $S = 2$ targets, it can be extended to an arbitrary number of targets as in [3].

4.2.5 DYNAMIC WAVEFORM SELECTION AND CONFIGURATION

The waveform selection algorithm seeks the waveform for each sensor that minimizes the sum of the predicted MSE for all targets at the next time step. The cost function to be minimized is still defined as in (2.19) but with the target state defined as in (4.9). The rectangular grid search algorithm of Chapter 3 is again used; however, the algorithm for approximating the predicted MSE consists of an extension to multiple targets of the algorithm presented in Section 4.1.4. This method of approximating the predicted cost is presented next.

Let $P_{k-1|k-1}$ represent the covariance of the state estimate at time $k-1$. We first obtain the predicted covariance at time k using the target dynamics model in (4.10) as

$$P_{k|k-1} = F P_{k-1|k-1} F^T + Q .$$

In order to obtain $P_{k|k}(\boldsymbol{\theta}_k)$, the estimate of the updated covariance that would result if a waveform parameterized by $\boldsymbol{\theta}_k$ was used at time k, we update each of the 4×4 block matrices $P^s_{k|k-1}$, along the diagonal of $P_{k-1|k-1}$, sequentially for each sensor.

Sigma points $\boldsymbol{\mathcal{X}}^s_n$ are first computed from the probability density function whose mean and covariance are given by the predicted position of target s and $P^s_{k|k-1}$, respectively, [29], as in (3.7). The sigma points are transformed to yield $\boldsymbol{\mathcal{Z}}^{s,i}_n = h^i(\boldsymbol{\mathcal{X}}^s_n)$. The covariance matrix $P^{s,i}_{zz}$ of $\boldsymbol{\mathcal{Z}}^{s,i}_n$, and the cross-covariance $P^{s,i}_{xz}$ between $\boldsymbol{\mathcal{X}}^s_n$ and $\boldsymbol{\mathcal{Z}}^{s,i}_n$ are computed analogously to (4.5) and (4.6), respectively. The updated covariance is computed as

$$P^{s,i}_{k|k}(\boldsymbol{\theta}^i_k) = P^s_{k|k-1} - q^{s,i}_{2k} P^c_{k|k} \qquad (4.15)$$

$$P^c_{k|k} = P^{s,i}_{xz}(P^{s,i}_{zz} + (N^s(\boldsymbol{\theta}^i_k))^{-1} P^{s,i}_{xz}{}^T \qquad (4.16)$$

where $q^{s,i}_{2k}$ is an information reduction factor [55]. The updates in (4.15) and (4.16) are performed for Sensor A to yield $P^{s,A}_{k|k}(\boldsymbol{\theta}^A_k)$. The update for Sensor B follows the same procedure but with $P^s_{k|k-1}$ in (4.15) replaced by $P^{s,A}_{k|k}(\boldsymbol{\theta}^A_k)$ to yield $P^s_{k|k}(\boldsymbol{\theta}_k) = P^{s,B}_{k|k}(\boldsymbol{\theta}^B_k)$. Note that the covariance matrices in (4.16) must be re-evaluated for the second update. Once each of the $P^s_{k|k}(\boldsymbol{\theta}_k)$ has been evaluated, we form $P_{k|k}(\boldsymbol{\theta}_k) = \text{diag}[P^1_{k|k}(\boldsymbol{\theta}_k), \ldots, P^S_{k|k}(\boldsymbol{\theta}_k)]$ and obtain $J(\boldsymbol{\theta}_k) \approx \text{trace}(\Lambda P_{k|k}(\boldsymbol{\theta}_k))$.

4.2.6   SIMULATION

In this simulation study, Sensor A and Sensor B were located at (3,000, -16,000) m and (3,000, 20,400) m, respectively, and tracked two targets as they moved in a two-dimensional space. The target trajectories were constructed with $\mathbf{x}^1_0 = [2,000 - 1,000 - 100\ 800]^T$ and $\mathbf{x}^2_0 = [1,000 - 1,000\ 100\ 800]^T$. In (2.10), $q = 1$ and the sampling interval was $\Delta T_s = 0.25$ s. The clutter density was $\rho = 0.0001$ false alarms per unit validation gate volume while the probability of false alarm was $P_f = 0.01$. The SNR was modeled as (4.12) with $r_0 = 50$ km. We used the standard particle filter as the target tracker, and the weighting matrix in (2.19) was the 8×8 identity matrix with appropriate units so that the MSE had units of m^2. Initially only Target 1 was present in the observation space and Target 2 was assumed to be detected at $k = 8$. Each target was tracked for 25 time steps (a total of 6.25 s). The initial position given to the tracker for target s was drawn from a normal distribution with mean $\mathbf{x}^s_0$ and variance $P_0 = \text{diag}[4,000, 4,000\ 100, 100]$. All results were averaged over 500 Monte Carlo runs, conditioned on convergence of the tracker. We defined convergence to imply that the target-originated measurements for each target fell outside their validation gates at either sensor no more than five times during the entire tracking sequence.

Figure 4.12 shows the averaged MSE obtained using the waveform selection and configuration algorithm. The abrupt increase in the MSE at $k = 8$ (1.75 (s) is due to the fact that the second target appears at this instant. Similarly, the drop in MSE at $k = 26$ (6.25 (s) is due to the fact that the first target disappears at this time. For comparison, we also show the averaged MSE obtained when the waveform duration and FM rate may be configured dynamically but the phase function is fixed. We observe that, with phase function agility, the tracking performance improves. In particular, the performance is significantly better than that achieved by the LFM chirp, which is a common choice in radar applications. A typical waveform selection during the tracking sequence is presented in Figure 4.13. Note that the waveforms selected include only the HFM and the PFM chirp with $\kappa = 3$, which corresponds to a parabolic instantaneous frequency. Interestingly, the LFM chirp is never selected and the selection behavior is similar to that in Figure 4.5.

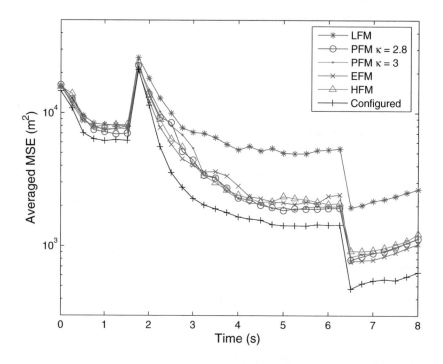

Figure 4.12: Averaged MSE using waveforms with fixed and agile phase functions.

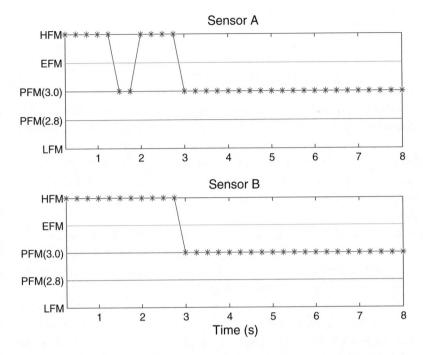

Figure 4.13: Typical waveform selection during the tracking sequence.

4.2.7   DISCUSSION

The behavior of the waveform selection algorithm as demonstrated in Figure 4.13, is due to the fact that, at the start of the tracking, the uncertainty in position and velocity are both large and the range and range-rate have to be estimated independently; thus waveforms with minimal range-Doppler coupling should be expected to be chosen. It has been demonstrated in Section 4.1.6, that the HFM chirp has this feature. When the uncertainty in the target state estimate is reduced, it is possible to obtain a reduction in estimation errors by exploiting the range-Doppler coupling. The power chirp with $\kappa = 3$ has been shown to provide a lower variance on range errors conditioned on range-rate than the other chirps while still minimizing the validation gate volume. Therefore, the algorithm selects this waveform during the later stages of the tracking sequence. Note that this behavior is repeated when the second target appears at time $k = 8$. At this time, the uncertainty in the state estimate again increases and the algorithm chooses the HFM chirp to independently estimate the range and range-rate of the second target.

CHAPTER 5

Conclusions

Recent advances in electromagnetics, processing, and radio frequency analog devices afford a new flexibility in waveform design and application. This technology is now being used in the research and development of sensors that can dynamically change the transmit or receive waveform in order to extract information that can contribute optimally to the overall sensing objective. Thus, a new paradigm of waveform-agile sensor processing has been exposed, which promises significant improvements in active sensor system performance by dynamic adaptation to the environment, target, or information to be extracted.

In this work, we presented new algorithms for waveform-agile sensing for target tracking. We considered various scenarios of this application, depending on the level of clutter embedded with the target. Specifically, we considered tracking in narrowband and wideband environments with no clutter, as well as with clutter, both for single and multiple targets.

Specifically, we considered the selection and configuration of waveforms for waveform-agile sensors so as to minimize a tracking error. In contrast with past research on this problem, we considered a nonlinear observations model to track a target moving in two dimensions as well as the use of waveforms with nonlinear time-frequency signatures. Our method of computing the predicted mean square error (MSE) is based on the unscented transform and the Cramér-Rao lower bound (CRLB) on the measurement error. To track a single target in clutter, we used a particle filter with probabilistic data association, while joint probabilistic data association was used in the tracking of multiple targets. The particle filter provides robustness with respect to observation nonlinearities and sensor positioning. The waveform selection algorithm was applied to several different scenarios including narrowband and wideband environments, clutter and imperfect detection, and single as well as multiple targets. For each case, we presented simulation studies to demonstrate that the algorithm provides approximately 10 dB reduction in MSE as compared to a nonadaptive system.

5.1 SUMMARY OF FINDINGS

A summary of the major conclusions we derived from our work is given below.

- **Waveforms that maximize the time-bandwidth product do not necessarily provide the best tracking performance.**
  The idea of using waveforms with the largest allowable duration and the maximum possible bandwidth is well established in radar. This configuration minimizes the *conditional* variance of range given range-rate estimates. However, such waveforms necessarily correlate the range and range-rate estimation errors and are not suitable when both these parameters have to be independently estimated. As we described in Chapter 4, when the tracking of a target

commences, neither parameter is known with adequate accuracy to achieve variance reduction by conditioning. Thus, waveforms that result in correlated errors in range and range-rate measurements are not appropriate. It is interesting to note that bats, which use sonar to locate and capture their prey, do not necessarily use waveforms that maximize the waveform time-bandwidth product. They vary the transmitted waveforms through various phases of the search, approach, and capture of their prey [59, 60]. That our algorithm behaves similarly to nature, provides in our opinion, a support of its validity.

- **Nonlinear frequency modulated (FM) waveforms offer significant advantages over the linear FM waveform.**
  The linear FM chirp has enjoyed widespread popularity in radar and sonar whereas nonlinear FM chirps have been relatively less used [32]. However, we have shown in Section 4.1.6 that the hyperbolic, exponential, and power-law FM chirp, with parabolic instantaneous frequency, provide better tracking performance than the linear FM chirp, especially when they are used in a dynamic, adaptive manner. Once again, the correspondence with nature is remarkable. Dolphins and bats are known to use nonlinear FM chirps and exploit their properties, such as Doppler tolerance [35, 59].

- **Dynamic waveform adaptation can reduce tracking error.**
  It has been recently shown [21] that for an LFM chirp waveform, the maximum information about the target state can be obtained by a waveform with a frequency sweep rate that is either the maximum or the minimum possible, given a set of constraints on the waveform such as average power, duration, and frequency sweep rates. This result was derived for a linear observations model and an information theoretic criterion. When the performance metric is the tracking MSE and when nonlinear observations models are used, we have demonstrated improved performance by dynamically selecting both the waveform duration and the frequency sweep rates of GFM chirps.

- **Dynamic waveform adaptation is more effective in the presence of clutter than in clutter-free environments.**
  When clutter is present in the observation space, there is uncertainty in the origin of the measurements. Accordingly, a validation gate must be set up and only those measurements that fall within the gate are considered as potentially target-originated. If clutter is assumed to be uniformly distributed, the number of false alarms is directly proportional to the validation gate volume. Since a large number of false alarms negatively impacts tracking performance, the control of the validation gate volume becomes important. The validation gate volume depends upon the choice of waveform as was shown in Section 4.1.6, and this provides an added dimension in which waveform adaptation can be exploited when used in the presence of clutter.

CRLB evaluation for Gaussian Envelope GFM Chirp from the ambiguity function

In this appendix, we derive the elements of the Fisher information matrix corresponding to GFM chirps (2.12) with a Gaussian envelope. The elements are obtained by computing the Hessian of the narrowband ambiguity function and evaluating it at the origin of the delay-Doppler plane.

As in (3.11), define the complex envelope as

$$\tilde{s}(t) = \left(\frac{1}{\pi\lambda^2}\right)^{\frac{1}{4}} \exp\left(-\frac{t^2}{2\lambda^2}\right) \exp\left(j2\pi b\xi(t)\right), \tag{A.1}$$

where λ parameterizes the duration of the Gaussian envelope, b is the frequency modulation (FM) rate, and $\xi(t)$ is a real valued, differentiable phase function. The narrowband ambiguity function is given by (2.18) and is repeated below for convenience:

$$AF_{\tilde{s}}(\tau, \nu) = \int_{-\infty}^{\infty} \tilde{s}\left(t + \frac{\tau}{2}\right)\tilde{s}^*\left(t - \frac{\tau}{2}\right)\exp(-j2\pi\nu t)dt. \tag{A.2}$$

Substituting (A.1) in (A.2)

$$AF_{\tilde{s}}(\tau, \nu) = \int_{-\infty}^{\infty} \frac{1}{\lambda\sqrt{\pi}}\exp\left\{-(1/\lambda^2)\left(t^2 + \tau^2/4\right)\right\}\exp\left\{j2\pi b\beta(\tau)\right\}\exp\left\{-j2\pi\nu t\right\}dt, \tag{A.3}$$

where

$$\beta(\tau) = \xi(t - \frac{\tau}{2}) - \xi(t + \frac{\tau}{2}).$$

Differentiating (A.3) with respect to τ,

$$\begin{aligned}
\frac{\partial AF_{\tilde{s}}(\tau, \nu)}{\partial \tau} &= \int_{-\infty}^{\infty} \frac{1}{\lambda\sqrt{\pi}}\exp\left\{-(1/\lambda^2)(t^2 + \tau^2/4)\right\} \cdot \\
&\quad \left[-\tau/(2\lambda^2) + j2\pi b\beta'(\tau)\right]\exp\left\{j2\pi b\beta(\tau)\right\} \cdot \\
&\quad \exp\left\{-j2\pi\nu t\right\}dt,
\end{aligned} \tag{A.4}$$

where

$$\beta'(\tau) = -\frac{1}{2}\left[\xi'(t - \frac{\tau}{2}) + \xi'(t + \frac{\tau}{2})\right]$$

and $\xi'(\tau) = d\xi(\tau)/d\tau$. Differentiating (A.4) with respect to τ,

$$
\frac{\partial^2 AF_{\tilde{s}}(\tau, \nu)}{\partial \tau^2} = \int_{-\infty}^{\infty} \frac{1}{\lambda\sqrt{\pi}} \exp\left\{-(1/\lambda^2)(t^2 + \tau^2/4)\right\} \cdot
$$
$$
\left[-1/(2\lambda^2) + (-\tau/(2\lambda^2))^2 + j2\pi b\beta''(\tau)\right.
$$
$$
\left. + j4\pi b(-\tau/(2\lambda^2))\beta'(\tau) + (j2\pi b\beta'(\tau))^2\right]
$$
$$
\exp\{j2\pi b\beta(\tau)\}\exp\{-j2\pi\nu t\}dt, \tag{A.5}
$$

where

$$
\beta''(\tau) = -\frac{1}{4}\left[-\xi''(t - \frac{\tau}{2}) + \xi''(t + \frac{\tau}{2})\right]
$$

and $\xi''(\tau) = d^2\xi(\tau)/d\tau^2$. Evaluating (A.5) at $\tau = 0$, $\nu = 0$,

$$
\left.\frac{\partial^2 AF_{\tilde{s}}(\tau, \nu)}{\partial \tau^2}\right|_{\substack{\tau=0 \\ \nu=0}} = -\frac{1}{2\lambda^2} - 4\pi^2 b^2 \cdot
$$
$$
\int_{-\infty}^{\infty} \frac{1}{\lambda\sqrt{\pi}} \exp\{t^2/\lambda^2\}(\xi'(t))^2 dt, \tag{A.6}
$$

where we have used $\beta''(0) = 0$ and $\beta'(0) = -\xi'(t)$.

Differentiating (A.3) twice with respect to ν,

$$
\frac{\partial^2 AF_{\tilde{s}}(\tau, \nu)}{\partial \nu^2} = \int_{-\infty}^{\infty} \frac{1}{\lambda\sqrt{\pi}} \exp\left\{-(1/\lambda^2)(t^2 + \tau^2/4)\right\} \cdot
$$
$$
(-j2\pi t)^2 \exp\{-j2\pi\nu t\}dt, \tag{A.7}
$$

and evaluating (A.7) at $\tau = 0$, $\nu = 0$,

$$
\left.\frac{\partial^2 AF_{\tilde{s}}(\tau, \nu)}{\partial \nu^2}\right|_{\substack{\tau=0 \\ \nu=0}} = -(2\pi)^2 \int_{-\infty}^{\infty} \frac{t^2}{\lambda\sqrt{\pi}} \exp\{-(t^2/\lambda^2)\}dt
$$
$$
= -(2\pi)^2 \frac{\lambda^2}{2}, \tag{A.8}
$$

where we have used $\beta(0) = 0$.

Differentiating (A.4) with respect to ν,

$$
\frac{\partial^2 AF_{\tilde{s}}(\tau, \nu)}{\partial \tau \partial \nu} = \int_{-\infty}^{\infty} \frac{1}{\lambda\sqrt{\pi}} \exp\left\{-(1/\lambda^2)(t^2 + \tau^2/4)\right\} \cdot
$$
$$
\left[-j2\pi t(1 + j2\pi b\beta'(\tau))\right]
$$
$$
\exp\{j2\pi b\beta(\tau)\}\exp\{-j2\pi\nu t\}dt. \tag{A.9}
$$

Evaluating (A.9) at $\tau = 0$, $\nu = 0$,

$$
\left.\frac{\partial^2 AF_{\tilde{s}}(\tau, \nu)}{\partial \tau \partial \nu}\right|_{\substack{\tau=0 \\ \nu=0}} = -2\pi \int_{-\infty}^{\infty} \frac{t\xi'(t)}{\lambda\sqrt{\pi}} \exp\{-(t^2/\lambda^2)\}dt. \tag{A.10}
$$

The elements of the Fisher Information matrix I, are the negative of the second derivatives of the ambiguity function, evaluated at $\tau = 0$, $\nu = 0$ [36]. From (A.6), (A.8) and (A.10),

$$I = \eta \begin{bmatrix} \frac{1}{2\lambda^2} + g(\xi) & 2\pi f(\xi) \\ 2\pi f(\xi) & (2\pi)^2 \frac{\lambda^2}{2} \end{bmatrix},$$

where, as in (3.12) and (3.13),

$$
\begin{aligned}
g(\xi) &= (2\pi b)^2 \int_{-\infty}^{\infty} \frac{[\xi'(t)]^2}{\lambda\sqrt{\pi}} \exp\left\{-\frac{t^2}{\lambda^2}\right\} dt, \\
f(\xi) &= 2\pi b \int_{-\infty}^{\infty} \frac{t\xi'(t)}{\lambda\sqrt{\pi}} \exp\left\{-\frac{t^2}{\lambda^2}\right\} dt.
\end{aligned}
$$

APPENDIX B

CRLB evaluation from the complex envelope

The elements of the Fisher information matrix can be computed directly from the complex envelope of the waveform, without explicitly evaluating the Hessian of the ambiguity function as in Appendix A. In this appendix, we provide expressions for these elements corresponding to the waveform with complex envelope defined by

$$\tilde{s}(t) = a(t) \exp{(j2\pi b\xi(t/t_r))}.$$

Let I denote the Fisher information matrix corresponding to the estimation of the delay and Doppler shift $[\tau \ \nu]^T$. For the narrowband case, the elements of I are given by [36]

$$
\begin{aligned}
I_{1,1} &= -\frac{\partial^2 AF_{\tilde{s}}(\tau, \nu)}{\partial \tau^2}\bigg|_{\substack{\tau=0 \\ \nu=0}} = \int_{-\lambda/2}^{\lambda/2} (\dot{a}^2(t) + a^2(t)\Omega^2(t))dt - \left[\int_{-\lambda/2}^{\lambda/2} a^2(t)\Omega(t)dt\right]^2 \\
I_{1,2} &= -\frac{\partial^2 AF_{\tilde{s}}(\tau, \nu)}{\partial \tau \partial \nu}\bigg|_{\substack{\tau=0 \\ \nu=0}} = \int_{-\lambda/2}^{\lambda/2} ta^2(t)\Omega^2(t)dt \\
I_{2,2} &= -\frac{\partial^2 AF_{\tilde{s}}(\tau, \nu)}{\partial \nu^2}\bigg|_{\substack{\tau=0 \\ \nu=0}} = \int_{-\lambda/2}^{\lambda/2} t^2 a^2(t)dt,
\end{aligned}
$$

where $\Omega(t) = 2\pi(b\frac{d}{dt}\xi(t/t_r) + f_c)$ and $I_{2,1} = I_{1,2}$. The Cramér-Rao lower bound (CRLB) for the measurement of $[r \ \dot{r}]$ at sensor i is then given by

$$N(\boldsymbol{\theta}^i) = \frac{1}{\eta}\Gamma I^{-1}\Gamma, \tag{B.1}$$

where $\Gamma = \mathrm{diag}[c/2, c/(2f_c)]$ and η is the SNR.

For the wideband case, $I_{1,1}$ is identical to that for the narrowband case even though the wideband ambiguity function is used to compute it. The other elements of I are [39]

$$
\begin{aligned}
I_{1,2} &= -\frac{\partial^2 WAF_{\tilde{s}}(\tau, \sigma)}{\partial \tau \partial \sigma}\bigg|_{\tau=0\sigma=1} = \int_{-\lambda/2}^{\lambda/2} t(\dot{a}^2(t) + a^2(t)\Omega^2(t))dt - \\
&\quad \int_{-\lambda/2}^{\lambda/2} a^2(t)\Omega(t)dt \cdot \int_{-\lambda/2}^{\lambda/2} ta^2(t)\Omega(t)dt
\end{aligned}
$$

$$I_{2,2} = -\frac{\partial^2 WAF_{\tilde{s}}(\tau, \sigma)}{\partial \sigma^2}\bigg|_{\tau=0\sigma=1} = \int_{-\lambda/2}^{\lambda/2} t^2(\dot{a}^2(t) + a^2(t)\Omega^2(t))dt - \left[\int_{-\lambda/2}^{\lambda/2} ta^2(t)\Omega(t)dt\right]^2 - \frac{1}{4}.$$

In this case also, the CRLB is given by (B.1) but with $\Gamma = \text{diag}[c/2, c/2]$.

APPENDIX C

Sample code for a simple particle filter

```
%-------------------------------------------------------------
% particle_filter.m
%-------------------------------------------------------------
% Tutorial particle filter application
%
% A particle filter that estimates the target state
% X_k = [x_k y_k x_dot_k y_dot_k]
% based on observations of its range and range rate from
% two independent radar/sonar sensors.
%
% Copyright Sandeep Sira 2008
%
%-------------------------------------------------------------

clear
close all

% Initialization
X_0 = [0 0 10 150]';              % Initial target state
a = [0 2000];                     % x-y coordinates of Sensor A
b = [3000 1500];                  % x-y coordinates of Sensor B
dT = 0.5;                         % Sampling interval in seconds
endk = 100;                       % Number of sampling epochs
R = diag([10 4 10 4]);            % Covariance matrix of sensor observation
                                  % error
q = 0.5;                          % Process noise intensity parameter
Ns = 500;                         % Number of particles
P_0 = diag([1000 100 1000 100]);  % Initial covariance

% State transition matrix
F = [1   0    dT   0;
     0   1    0    dT;
     0   0    1    0;
     0   0    0    1];

% Process noise covariance matrix
Q = q*[ dT^3/3   0      dT^2/2   0;
        0        dT^3/3 0        dT^2/2;
        dT^2/2   0      dT       0;
        0        dT^2/2 0        dT];

% Generate a sample trajectory
X = X_0;
```

```
sqrtm_Q = sqrtm(Q);
for k = 2:endk
    X(:,k) = F*X(:,k-1) + sqrtm_Q*randn(4,1);
end

% Sample an initial value of the target state.
% This is the mean of the  density P(X_0)
XHat(:,1) = X_0 + sqrtm(P_0)*randn(4,1);

% Sample the particles from the initial density
Xp = XHat*ones(1,Ns) + sqrtm(P_0)*randn(4,Ns);

% Commence particle filtering loop
sqrtm_R = sqrtm(R);
inv_R = inv(R);
for k = 2:endk
    % Project the particles forward according to the target kinematics
    % model
    Xp = F*Xp + sqrtm_Q*randn(4, Ns);

    % Get the true observation
    zeta = X(:,k);
    % Range from Sensor A
    da_true = norm( zeta([1:2]) - a');
    % Range from Sensor B
    db_true =  norm( zeta([1:2]) - b');
    % Range-rate wrt Sensor A
    va_true = zeta([3:4])' * (zeta([1:2]) - a')/da_true;
    % Range-rate wrt Sensor B
    vb_true = zeta([3:4])' * (zeta([1:2]) - b')/db_true;

    % Add the sensor noise
    Zk = [ da_true; va_true; db_true; vb_true] + sqrtm_R*randn(4,1);

    % Obtain the observations corresponding to each particle
    da = sqrt( sum( (Xp([1,2],:) - a'*ones(1,Ns)).^ 2 ) );
    db = sqrt( sum( (Xp([1,2],:) - b'*ones(1,Ns)).^ 2 ) );
    va = diag(Xp([3:4],:)' * ( Xp([1:2],:) - a'*ones(1,Ns)))' ./ da;
    vb = diag(Xp([3:4],:)' * ( Xp([1:2],:) - b'*ones(1,Ns)))' ./ db;

    % Obtain a matrix of range and range-rates, one column per particle
    Z = [da;  va;  db;  vb];

    % Calculate the weights
    innov = Zk*ones(1,Ns)  -  Z;

    w = diag( exp(-0.5 * innov' * inv_R * innov) )';

    % Normalize the weights
    if sum(w) == 0
        display('Alarm weights = 0. Setting all weights equal!')
        w = ones(1,Ns)/Ns;
    end
    w = w / sum(w);

    % Calculate the estimate
```

```
    XHat(:,k) = Xp*w';

    % Plot the results so far
    figure(1);  clf;
    plot(X(1,:), X(2,:), 'b*-', XHat(1,:), XHat(2,:), 'co-', ...
            Xp(1,:), Xp(2,:), 'r.');
    hold on
    %plot(a(1), a(2), 'rs', b(1), b(2), 'gs');
    title('X-Y position');  xlabel('X (m)');     ylabel('Y (m)');
    legend('Target trajectory', 'Target estimate', ...
                'Particles', 'Location', 'SouthEast')

    figure(2);  clf;
    subplot(2,1,1),
    plot([0:endk-1]*dT, X(3,:), 'b*-', [0:k-1]*dT, XHat(3,:),   ...
            'co-', (k-1)*dT*ones(1,Ns), Xp(3,:), 'r.');
    title('Velocity');  ylabel('X m/s');
    subplot(2,1,2),
    plot([0:endk-1]*dT, X(4,:), 'b*-', [0:k-1]*dT, XHat(4,:),   ...
            'co-', (k-1)*dT*ones(1,Ns), Xp(4,:), 'r.');
    xlabel('Sampling instant');        ylabel('Y m/s');

    % Resample the particles
    % Resample
    Xp = importance_resample(Xp, w);

end

% Plot the MSE
figure
trMSE = diag((XHat - X)' * (XHat - X));
semilogy([0:endk-1]*dT, trMSE, 'b*-');
xlabel('Time'); ylabel('MSE');
title('Total Mean Square Tracking Error');

% -------------------------------------------------------------------

function [Xpout, wout, index] = importance_resample(Xp, w)

% Carry out resampling.
% Returns Particles, weights and index of parent of each particle

[a M] = size(Xp);

% Cumulative Distributive Function
cumpr = cumsum(w(1,1:M))';
cumpr = cumpr/max(cumpr);

u(1,1) = (1/M)*rand(1,1);
i=1;
for j = 1:M
    u(j,1)= u(1,1) + (1/M)*(j-1);
    while (u(j,1) > cumpr(i,1))
        i = i+1;
        if i > M
            break
```

```
        end
    end
    if i ≤ M
        x_update(:,j) = Xp(:,i);
        index(j) = i;
    end
end

Xpout = x_update;
wout = 1/M*ones(1,M);

return
```

APPENDIX D

Sample code for an unscented particle filter

```
%-----------------------------------------------------------
% unscented_particle_filter.m
%-----------------------------------------------------------
% Unscented particle filter application
%
% A particle filter based on the unscented transform that
% estimates the target state X_k = [x_k y_k x_dot_k y_dot_k]
% based on observations of its range and range rate from
% two independent radar/sonar sensors.
%
% Copyright Sandeep Sira 2008
%
%-----------------------------------------------------------
clear
close all

% Initialization
X_0 = [0 0 10 150]';              % Initial target state
a = [0 2000];                     % x-y coordinates of Sensor A
b = [3000 1500];                  % x-y coordinates of Sensor B
dT = 0.5;                         % Sampling interval in seconds
endk = 100;                       % Number of sampling epochs
R = diag([10 4 10 4]);            % Covariance matrix of sensor observation
                                  % error
q = 0.5;                          % Process noise intensity parameter
Ns = 10;                          % Number of particles
P_0 = diag([1000 100 1000 100]);  % Initial covariance

% UPF parameters
na = 4;     kappa = 4;   alpha = 1;  beta = 0;
lambda = alpha^2*(na+kappa) - na;

% State transition matrix
F = [1   0    dT   0;
     0   1    0    dT;
     0   0    1    0;
     0   0    0    1];

% Process noise covariance matrix
Q = q*[ dT^3/3    0     dT^2/2   0;
        0      dT^3/3   0      dT^2/2;
        dT^2/2    0     dT       0;
        0      dT^2/2   0      dT];
```

```matlab
sqrtm_Q = sqrtm(Q);
inv_Q = inv(Q);
inv_R = inv(R);
sqrtm_R = sqrtm(R);

% Generate a target trajectory
X = X_0;
for k = 2:endk
    X(:,k) = F*X(:,k-1) + sqrtm_Q*randn(4,1);
end

% Sample an initial value of the target state. This is the mean of the
% density P(X_0)
XHat(:,1) = X_0 + sqrtm(P_0)*randn(4,1);

% Sample the particles from the initial density
Xp = XHat*ones(1,Ns) + sqrtm(P_0)*randn(4,Ns);

% Mean and covariance of particles
mx = mean(Xp,2)*ones(1,Ns);     err_X = Xp - mx;
for j = 1:Ns
    P_xx(:,:,j) = err_X(:,j) * err_X(:,j)';
end

% Commence unscented particle filtering loop
for k = 2:endk
    % Project the particles forward
    Xp_in = F*Xp;

    % Get the observation
    zeta = X(:,k);
    da_true = norm( zeta([1:2]) - a');
    db_true =  norm( zeta([1:2]) - b');
    va_true = zeta([3:4])' * (zeta([1:2]) - a')/da_true;
    vb_true = zeta([3:4])' * (zeta([1:2]) - b')/db_true;
    Zk = [ da_true; va_true; db_true; vb_true] + sqrtm_R*randn(4,1);

    for j = 1:Ns

        % Obtain the sigma points
        P(:,:) = P_xx(:,:,j);
        sqrt_P_xx = real(sqrtm( (na + lambda)*P ));
        chi_X(:,:) = mx(:,j)*ones(1,2*na+1) + ...
            [ zeros(4,1)   sqrt_P_xx   -sqrt_P_xx ];
        w_chi = [ kappa ones(1,2*na)/2 ] / (na + kappa);

        % Propagate the sigma points
        chi_X = F*chi_X + sqrtm_Q*randn(4,2*na+1);

        % Calculate mean and variance
        mx(:,j) = chi_X * w_chi';
        err_chi_X = chi_X - mx(:,j)*ones(1,2*na+1);
        PHat = err_chi_X* (err_chi_X.*(ones(4,1)*w_chi))';

        % Range and range rate of the sigma points as
```

```
% observed by the sensors
chi_X_da = sqrt( sum( (chi_X([1,2],:) - a'*ones(1,9)).^ 2 ) );
chi_X_db = sqrt( sum( (chi_X([1,2],:) - b'*ones(1,9)).^ 2 ) );
chi_X_va = diag(chi_X([3:4],:)' ...
    * ( chi_X([1:2],:) - a'*ones(1,9)))' ./ chi_X_da;
chi_X_vb = diag(chi_X([3:4],:)' ...
    * ( chi_X([1:2],:) - b'*ones(1,9)))' ./ chi_X_db;

chi_Z = [chi_X_da;  chi_X_va;  chi_X_db;  chi_X_vb];
m_chi_Z = chi_Z * w_chi';
err_chi_Z = chi_Z - m_chi_Z*ones(1,9);

% Obtain P_zz, P_xz
P_zz = zeros(4,4);        P_xz = zeros(4,4);
P_zz = err_chi_Z * (err_chi_Z.*(ones(4,1)*w_chi))';
P_xz = err_chi_X * (err_chi_Z.*(ones(4,1)*w_chi))';

% Innovations covariance and Kalman gain
P_eta = P_zz + R;      K = P_xz * inv(P_eta);

% Update
mx(:,j) = mx(:,j) + K * (Zk - m_chi_Z);
PHat = PHat - K * P_eta * K';

% Store for the next time step
P_xx(:,:,j) = PHat;

% Sample a new particle
Xp(:,j) = mx(:,j) + real(sqrtm(PHat))*randn(4,1);

% Importance density component of the weight
w3(j) = exp(-0.5 * ( Xp(:,j) - mx(:,j) )' ...
    * inv(PHat) * ( Xp(:,j) - mx(:,j) ) );

end


% Calculate the weights
da = sqrt( sum( (Xp([1:2],:) - a'*ones(1,Ns)).^2 ));
db = sqrt( sum( (Xp([1:2],:) - b'*ones(1,Ns)).^2 ));
va = diag(Xp([3:4],:)' * ( Xp([1:2],:) - a'*ones(1,Ns)))' ./ da;
vb = diag(Xp([3:4],:)' * ( Xp([1:2],:) - b'*ones(1,Ns)))' ./ db;

innov = Zk*ones(1,Ns)  - [ da; va; db; vb];

% Likelihood component
w1 = diag( exp(-0.5 * innov' * inv_R * innov) )';
% Kinematic prior component
w2 = diag( exp(-0.5 * ( Xp - Xp_in )' * inv_Q * ...
    ( Xp - Xp_in ) ) )';

w = w1 .* w2 ./ w3;

if sum(w) == 0
    fprintf('Alarm weights = 0 at time step %d\n', k);
    w = ones(1,Ns)/Ns;
end
```

```
    w = w / sum(w);

    N_eff = 1 / (w * w');

    % Calculate the estimate
    XHat(:,k) = Xp*w';

    % Plot the results so far
    figure(1);  clf;
    plot(X(1,:), X(2,:), 'b*-', XHat(1,:), XHat(2,:), 'co-', ...
         Xp(1,:), Xp(2,:), 'r.');
    hold on
    %plot(a(1), a(2), 'rs', b(1), b(2), 'gs');
    title('X-Y position');  xlabel('X (m)');    ylabel('Y (m)');
    legend('Target trajectory', 'Target estimate', ...
           'Particles', 'Location', 'SouthEast')

    figure(2);  clf;
    subplot(2,1,1),
    plot([0:endk-1]*dT, X(3,:), 'b*-', [0:k-1]*dT, XHat(3,:), ...
         'co-', (k-1)*dT*ones(1,Ns), Xp(3,:), 'r.');
    title('Velocity');  ylabel('X m/s');
    subplot(2,1,2),
    plot([0:endk-1]*dT, X(4,:), 'b*-', [0:k-1]*dT, XHat(4,:), ...
         'co-', (k-1)*dT*ones(1,Ns), Xp(4,:), 'r.');
    xlabel('Sampling instant');    ylabel('Y m/s');

    % Resample
    [Xp w index] = importance_resample(Xp, w);

    % Housekeeping
    mx = mx(:,index);
    P_xx = P_xx(:,:,index);

end

fprintf('\n\n');

% Plot the MSE
figure
trMSE = diag((XHat - X)' * (XHat - X));
semilogy([0:endk-1]*dT, trMSE, 'b*-');
xlabel('Time'); ylabel('MSE');
title('Total Mean Square Tracking Error');
```

Bibliography

[1] A. Drozd, "Waveform diversity and design," IEEE EMC Society Newsletter, Summer 2006. [Online]. Available: http://www.ieee.org/organizations/pubs/newsletters/emcs/summer06/cover_story.pdf

[2] D. Cochran, "Reducing complexity for defense," *DARPATech 2002 Symposium*, Jul. 2002. [Online]. Available: http://web-ext2.darpa.mil/DARPATech2002/presentations/dso_pdf/speeches/COCHRAN.pdf

[3] Y. Bar-Shalom and T. E. Fortmann, *Tracking and Data Association*. Boston: Academic Press, 1988.

[4] R. L. Mitchell and A. W. Rihaczek, "Matched filter responses of the linear FM waveform," *IEEE Trans. Aerosp. Electron. Syst.*, vol. 4, pp. 417–432, May 1968.

[5] D. F. DeLong and E. M. Hofstetter, "On the design of optimum radar waveforms for clutter rejection," *IEEE Trans. Inform. Theory*, vol. IT-13, no. 3, pp. 454–463, Jul. 1967. DOI: 10.1109/TIT.1967.1054038

[6] R. McAulay and J. Johnson, "Optimal mismatched filter design for radar ranging, detection, and resolution," *IEEE Trans. Inform. Theory*, vol. 17, pp. 696–701, Nov. 1971. DOI: 10.1109/TIT.1971.1054722

[7] M. Athans and F. C. Schweppe, "Optimal waveform design via control theoretic principles," *Information and Control*, vol. 10, pp. 335–377, Apr. 1967. DOI: 10.1016/S0019-9958(67)90183-0

[8] F. C. Schweppe and D. L. Gray, "Radar signal design subject to simultaneous peak and average power constraints," *IEEE Trans. Inform. Theory*, vol. IT-12, pp. 13–26, Jan. 1966. DOI: 10.1109/TIT.1966.1053853

[9] M. R. Bell, "Information theory and radar waveform design," *IEEE Trans. Inform. Theory*, vol. 39, pp. 1578–1597, Sep. 1993. DOI: 10.1109/18.259642

[10] K. T. Wong, "Adaptive pulse-diverse radar/sonar waveform design," in *IEEE Radar Conference*, pp. 105–110, May 1998. DOI: 10.1109/NRC.1998.677985

[11] N. Wang, Y. Zhang, and S. Wu, "Radar waveform design and target detection using wavelets," in *CIE International Conference on Radar*, pp. 506–509, Oct. 2001. DOI: 10.1109/ICR.2001.984758

[12] R. J. Bonneau, "A wavelet packet basis optimization approach to radar waveform design," in *IEEE International Symposium on Antennas and Propagation*, vol. 4, pp. 814–816, Jul. 2001. DOI: 10.1109/APS.2001.959589

[13] S. M. Sowelam and A. H. Tewfik, "Waveform selection in radar target classification," *IEEE Trans. Inform. Theory*, vol. 46, pp. 1014–1029, May 2000. DOI: 10.1109/18.841178

[14] D. Garren, M. Osborn, A. Odom, J. Goldstein, S. Pillai, and J. Guerci, "Enhanced target detection and identification via optimised radar transmission pulse shape," *IEE Proceedings-Radar Sonar and Navigation*, vol. 148, no. 3, pp. 130–138, 2001. DOI: 10.1049/ip-rsn:20010324

[15] S.-M. Hong, R. J. Evans, and H.-S. Shin, "Control of waveforms and detection thresholds for optimal target tracking in clutter," in *Proceedings of the 39th IEEE Conference on Decision and Control*, vol. 4, pp. 3906–3907, Dec. 2000. DOI: 10.1109/CDC.2000.912323

[16] S.-M. Hong, R. J. Evans, and H.-S. Shin, "Optimization of waveform and detection threshold for target tracking in clutter," in *Proceedings of the 40th SICE Annual Conference*, pp. 42–47, Jul. 2005. DOI: 10.1109/SICE.2001.977803

[17] D. J. Kershaw and R. J. Evans, "Optimal waveform selection for tracking systems," *IEEE Trans. Inform. Theory*, vol. 40, no. 5, pp. 1536–1550, Sep. 1994. DOI: 10.1109/18.333866

[18] D. J. Kershaw and R. J. Evans, "Waveform selective probabilistic data association," *IEEE Trans. Aerosp. Electron. Syst.*, vol. 33, pp. 1180–1188, Oct. 1997. DOI: 10.1109/7.625110

[19] C. Rago, P. Willett, and Y. Bar-Shalom, "Detection-tracking performance with combined waveforms," *IEEE Trans. Aerosp. Electron. Syst.*, vol. 34, pp. 612–624, Apr. 1998. DOI: 10.1109/7.670395

[20] R. Niu, P. Willett, and Y. Bar-Shalom, "Tracking considerations in selection of radar waveform for range and range-rate measurements," *IEEE Trans. Aerosp. Electron. Syst.*, vol. 38, no. 2, pp. 467–487, Apr. 2002. DOI: 10.1109/TAES.2002.1008980

[21] S. D. Howard, S. Suvorova, and W. Moran, "Waveform libraries for radar tracking applications," *International Conference on Waveform Diversity and Design*, Edinburgh, UK, Nov. 2004. DOI: 10.1109/CISS.2006.286688

[22] S. P. Sira, A. Papandreou-Suppappola, and D. Morrell, "Dynamic configuration of time-varying waveforms for agile sensing and tracking in clutter," *IEEE Trans. Signal Processing*, vol. 55, no. 1, pp. 3207–3217, Jun. 2007. DOI: 10.1109/TSP.2007.894418

[23] S. M. Hong, R. J. Evans, and H. S. Shin, "Optimization of waveform and detection threshold for range and range-rate tracking in clutter," *IEEE Trans. Aerosp. Electron. Syst.*, vol. 41, no. 1, pp. 17–33, Jan. 2005. DOI: 10.1109/TAES.2005.1413743

[24] B. F. L. Scala, W. Moran, and R. J. Evans, "Optimal adaptive waveform selection for target detection," in *International Conference on Radar*, pp. 492–496, 2003. DOI: 10.1109/RADAR.2003.1278791

[25] A. Doucet, N. de Freitas, and N. Gordon, Eds., *Sequential Monte Carlo Methods in Practice*. Springer-Verlag, 2001.

[26] M. S. Arulampalam, S. Maskell, N. Gordon, and T. Clapp, "A tutorial on particle filters for online nonlinear/non-Gaussian Bayesian tracking," *IEEE Trans. Signal Processing*, vol. 50, pp. 174–188, Feb. 2002. DOI: 10.1109/78.978374

[27] E. Kalman, Rudolph, "A new approach to linear filtering and prediction problems," *Transactions of the ASME–Journal of Basic Engineering*, vol. 82, no. Series D, pp. 35–45, 1960.

[28] R. van der Merwe, N. de Freitas, A. Doucet, and E. Wan, "The Unscented Particle Filter," in *Advances in Neural Information Processing Systems 13*, Nov. 2001.

[29] S. Julier and J. Uhlmann, "A new extension of the Kalman filter to nonlinear systems," *International Symposium on Aerospace/Defense Sensing, Simulation and Controls*, 1997. DOI: 10.1117/12.280797

[30] L. J. Ziomek, *Underwater Acoustics: A Linear Systems Theory Approach*.Orlando, FL: Academic Press, 1985.

[31] R. O. Nielsen, *Sonar Signal Processing*. Artech House, 1991.

[32] M. I. Skolnik, *Radar Handbook*, 2nd ed., M. I. Skolnik, Ed. New York: McGraw Hill, 1990.

[33] A. Papandreou-Suppappola, R. L. Murray, B. G. Iem, and G. F. Boudreaux-Bartels, "Group delay shift covariant quadratic time-frequency representations," *IEEE Trans. Signal Processing*, vol. 49, pp. 2549–2564, 2001. DOI: 10.1109/78.960403

[34] A. Papandreou-Suppappola, *Time-Varying Processing: Tutorial on Principles and Practice in Applications in Time-Frequency Signal Processing*, A. Papandreou-Suppappola, Ed. Florida: CRC Press, 2002.

[35] R. A. Altes and E. L. Titlebaum, "Bat signals as optimally Doppler tolerant waveforms," *Journal of the Acoustical Society of America*, vol. 48, pp. 1014–1020, Oct. 1970. DOI: 10.1121/1.1912222

[36] H. L. Van Trees, *Detection Estimation and Modulation Theory, Part III*. New York: Wiley, 1971.

[37] L. G. Weiss, "Wavelets and wideband correlation processing," *IEEE Signal Processing Mag.*, pp. 13–32, Jan. 1994. DOI: 10.1109/79.252866

[38] L. Cohen, *Time-Frequency Analysis.*   Englewood Cliffs, New Jersey: Prentice-Hall, 1995.

[39] D. W. Ricker, *Echo Signal Processing.*   Kluwer Academic Publishers, 2003.

[40] P. L'Ecuyer, "An overview of derivative estimation," *Proceedings of the 1991 Winter Simulation Conference*, pp. 207–217, Dec. 1991. DOI: 10.1109/WSC.1991.185617

[41] G. Pflug, *Optimization of Stochastic Models: The Interface Between Simulation and Optimization.* Kluwer Academic Publishers, 1996.

[42] J. C. Spall, "Multivariate stochastic approximation using a simultaneous perturbation gradient approximation," *IEEE Trans. Automat. Contr.*, vol. 37, pp. 332–341, Mar. 1992. DOI: 10.1109/9.119632

[43] J. Kiefer and J. Wolfowitz, "Stochastic estimation of a regression function," *Annals of Mathematical Statistics*, vol. 23, pp. 462–466, Sep. 1952. DOI: 10.1214/aoms/1177729392

[44] P. Sadegh and J. C. Spall, "Constrained optimization via stochastic approximation with a simultaneous perturbation gradient approximation," *Automatica*, vol. 33, no. 5, pp. 889–892, 1997. DOI: 10.1016/S0005-1098(96)00230-0

[45] S. P. Sira, D. Morrell, and A. Papandreou-Suppappola, "Waveform design and scheduling for agile sensors for target tracking," *Asilomar Conference on Signals, Systems and Computers*, vol. 1, pp. 820–824, Nov. 2004. DOI: 10.1109/ACSSC.2004.1399251

[46] B. D. Anderson and J. B. Moore, *Optimal Filtering.*   New Jersey: Prentice-Hall, 1979.

[47] S. P. Sira, A. Papandreou-Suppappola, and D. Morrell, "Time-varying waveform selection and configuration for agile sensors in tracking applications," *IEEE International Conference on Acoustics, Speech, and Signal Processing*, vol. 5, pp. 881–884, Mar. 2005. DOI: 10.1109/ICASSP.2005.1416445

[48] D. A. Swick, "A review of wideband ambiguity functions," Naval Research Laboratory, Washington, D.C., Tech. Rep. 6994, Dec. 1969.

[49] Q. Jin, K. M. Wong, and Z.-Q. Luo, "The estimation of time delay and doppler stretch of wideband signals," *IEEE Trans. Signal Processing*, vol. 43, no. 4, pp. 904–916, Apr. 1995. DOI: 10.1109/78.376843

[50] Y. Doisy, L. Deruaz, S. P. Beerens, and R. Been, "Target doppler estimation using wideband frequency modulated signals," *IEEE Trans. Signal Processing*, vol. 48, no. 5, pp. 1213–1224, May 2000. DOI: 10.1109/78.839970

[51] C. E. Cook and M. Bernfeld, *Radar Signals, An Introduction to Theory and Application.*   Boston: Artech House, 1993.

[52] S. P. Sira, A. Papandreou-Suppappola, and D. Morrell, "Waveform scheduling in wideband environments," *IEEE International Conference on Acoustics, Speech, and Signal Processing*, vol. 1, pp. I–697 – I–700, May 2006. DOI: 10.1109/ICASSP.2006.1661477

[53] S. P. Sira, A. Papandreou-Suppappola, and D. Morrell, "Characterization of waveform performance in clutter for dynamically configured sensor systems," *Waveform Diversity and Design Conference*, Lihue, Hawaii, Jan. 2006.

[54] R. Niu, P. Willett, and Y. Bar-Shalom, "Matrix CRLB scaling due to measurements of uncertain origin," *IEEE Trans. Signal Processing*, vol. 49, no. 7, pp. 1325–1335, Jul. 2001. DOI: 10.1109/78.928687

[55] T. Fortmann, Y. Bar-Shalom, M. Scheffe, and S. Gelfand, "Detection thresholds for tracking in clutter - A connection between estimation and signal processing," *IEEE Trans. Automat. Contr.*, vol. 30, pp. 221–229, Mar. 1985. DOI: 10.1109/TAC.1985.1103935

[56] D. J. Kershaw and R. J. Evans, "A contribution to performance prediction for probabilistic data association tracking filters," *IEEE Trans. Aerosp. Electron. Syst.*, vol. 32, no. 3, pp. 1143–1147, Jul. 1996. DOI: 10.1109/7.532274

[57] S. P. Sira, A. Papandreou-Suppappola, and D. Morrell, "Waveform-agile sensing for tracking multiple targets in clutter," *Conference on Information Sciences and Systems*, Princeton, NJ, pp. 1418–1423, Mar. 2006. DOI: 10.1109/CISS.2006.286687

[58] C. Kreucher, K. Kastella, and Alfred O. Hero, III, "Multitarget tracking using the joint multitarget probability density," *IEEE Trans. Aerosp. Electron. Syst.*, vol. 41, no. 4, pp. 1396–1414, Oct. 2005. DOI: 10.1109/TAES.2005.1561892

[59] J. A. Simmons and R. Stein, "Acoustic imaging in bat sonar: Echolocation signals and the evolution of echolocation," *Journal of Comparative Psychology*, vol. 135, no. 1, pp. 61–84, Mar. 1980. DOI: 10.1007/BF00660182

[60] J. A. Simmons, P. A. Saillant, and S. P. Dear, "Through a bat's ear," *IEEE Spectrum*, vol. 29, pp. 46–48, Mar. 1992. DOI: 10.1109/6.123330

Biography

Sandeep P. Sira received the M.Tech. degree from the Indian Institute of Technology, Kanpur, India, in 1999, and the Ph.D. degree in electrical engineering in 2007 from Arizona State University (ASU). In 2007, he was a post-doctoral research associate at ASU. He is currently with Zounds Inc., in Mesa, Arizona. His research interests include waveform-agile sensing, target tracking, audio signal processing, and detection and estimation theory.

Antonia Papandreou-Suppappola is a Professor in the Department of Electrical Engineering at Arizona State University (ASU). She is the co-Director of the Sensor, Signal and Information Processing (SenSIP) Center, and Associate Director of the Adaptive Intelligent Materials and Systems (AIMS) Center at ASU. Her research interests are in the areas of Waveform-Agile Sensing, Time-Frequency Signal Processing, and Statistical Signal Processing. She is currently funded by an Air Force Office of Scientific Research (AFOSR) Multidisciplinary University Research Initiative (MURI) on Adaptive Waveform Design for Full Spectral Dominance, and she participated in the DARPA Waveform-Agile Sensing, program, where she worked on waveform selection algorithms for sensing in highly cluttered environments.

Darryl Morrell is an associate professor of engineering at Arizona State University, where he is participating in the design and implementation of a multidisciplinary undergraduate engineering program using innovative, research-based pedagogical and curricular approaches. His research interests include stochastic decision theory applied to sensor scheduling and information fusion as well as application of research-based pedagogy to engineering education.

Printed in the United States
by Baker & Taylor Publisher Services